Signals and Systems

Made Ridiculously Simple

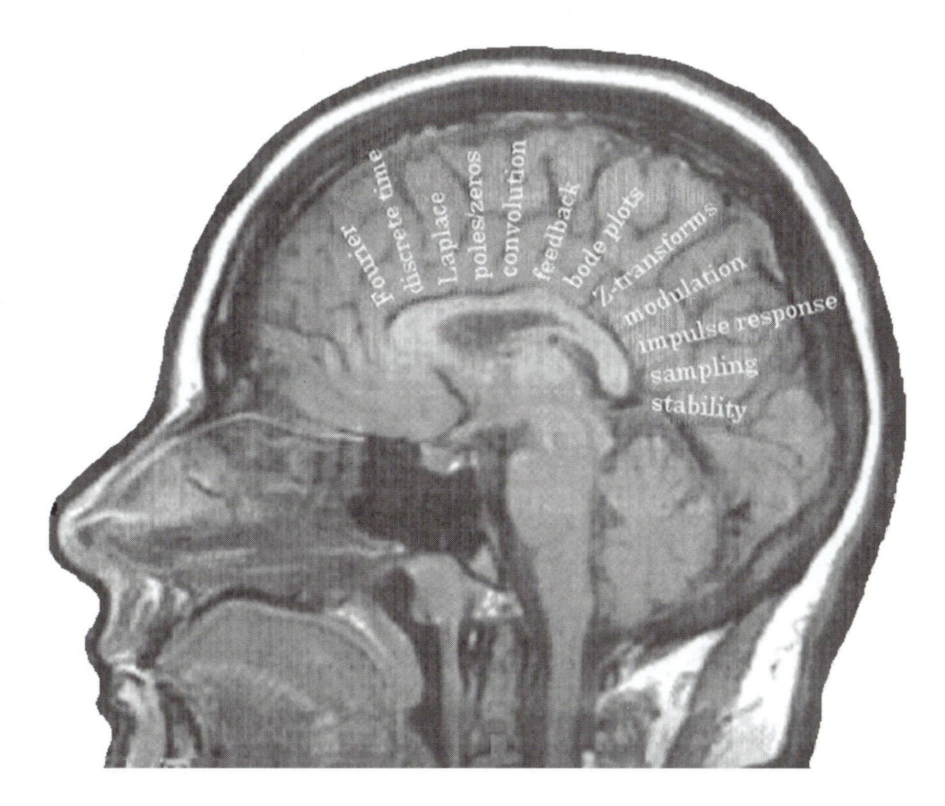

Zoher Z. Karu

ZiZi Press
Cambridge, MA

Fourth Printing

Printed in the United States of America

Publisher's Cataloging-in-Publication Data

Library of Congress Catalog Card Number: 94–90623

ISBN 0-9643752-1-4

Includes bibliographical references.

Keywords: 1. Signal theory (telecom). 2. System analysis. 3. Signal processing. I. Title.

ZiZi Press
1404 Old Carriage Lane
Huntsville, AL 35802 USA

Tel: (256) 520-5249
FAX: (256) 883-3124
E-mail: books@zizipress.com

World Wide Web: http://www.zizipress.com

To new beginnings

"Never let schooling interfere with your education."

- Mark Twain

"Be the master of your fate, not its slave."

- Zoher Z. Karu

Success

To laugh often and much;
to win the respect of intelligent people and affection of children;
to earn the appreciation of honest critics and endure the betrayal of false friends;
to appreciate beauty, to find the best in others;
to leave the world a bit better, whether by a healthy child, a garden patch or a redeemed social condition;
to know even one life has breathed easier because you have lived.
This is to have succeeded.

- Ralph Waldo Emerson

Preface

The origins of this book date back to the Fall of 1992 when I was appointed to be the head teaching assistant for 6.003, a semester-long core electrical engineering class at M.I.T. called "Signals and Systems." During the term, I put together a small handwritten review packet to help my students prepare for the midterm exam. I decided to call it "6.003 Made Ridiculously Simple." The notes were an instant success, and many people commented on their usefulness to me. Although I was no longer teaching, the same packet became a regular handout for the next several semesters. The apparent popularity and need for such a set of notes prompted me to write this book.

Traditional textbooks in this field often leave the reader to guess what's important and what's not, often losing track of "the big picture." This book contains, in my opinion, the fundamental concepts behind signal processing and linear system theory. It tells you what you need to know and tells it to you fast. This book is not meant to replace a textbook, but rather supplement one. It is designed to be used in two ways: as a study guide while taking a signal processing or linear systems course and as a reference book for rapidly reviewing the material for an exam, the Ph.D. qualifiers, or before taking a more advanced course in this area.

The book is written from a student's perspective, providing practical advice on problem-solving skills while detailing areas that, in my experience, have been known to give people difficulty. Each chapter can be considered to be a set of lecture or tutorial notes on that topic. The appendix of this book is a compilation of a variety of mathematical concepts that professors often view as assumed knowledge, but students can't seem to find written down anywhere.

I have intentionally not included several pages of exercises at the end of each chapter – there are plenty of textbooks for that purpose. Rather, each chapter contains carefully selected examples and sample problems that reinforce the fundamental issues on that topic. My philosophy is that overly complicated problems fail to verify the core concepts in the student's mind; in that case, students tend to learn by "pattern-matching" rather than truly trying to understand the material.

The book is based on a one-semester M.I.T. Electrical Engineering course, which runs at the typical "like trying to take a drink from a fire hose" pace. There is enough material to accompany a more in-depth two semester course if deemed necessary. Also, although many of the examples are circuit-based, the material in this book should appeal to several different engineering disciplines.

A great deal of effort was put into organizing the material in this book. The best part about signals and systems is that everything fits together so nicely. Although this may be intellectually satisfying when you're done, it makes this subject very difficult to teach since everything is so interconnected. The reader will get the most out of this book and subject matter by taking the time to identify and appreciate the unifying concepts present.

As part of my first venture as an author, I decided to start my own publishing company to produce this book. This of course means that the entire burden of writing, proofreading, editing, designing, and printing rests solely on my shoulders. Believe me, it takes more time than you might expect. If there is one thing I've learned, it's that I have great respect for people who can find the time to write a 700 page textbook.

As with any work of this magnitude, there are several people that have made this book possible. I would first like to thank Professor William Siebert for giving me the opportunity to be Head TA for 6.003; it was truly an enjoyable experience. I would also like to thank Deron Jackson, who is by now a 6.003 teaching machine. The time Deron spent proofreading and providing suggestions for my chapters has been invaluable. He also provided me with several FrameMaker and MATLAB tips, including a very useful utility he wrote for importing MATLAB graphs into FrameMaker. Elmer Hung, Rajeev Surati, Ed Chalom, and Angela Hsieh also provided many much-needed proofreading comments and managed to catch a few last-minute errors. I would also like to thank all of the 6.003 students who have both learned from and have given me feedback on these notes as they evolved into a book. Finally, I have to thank my parents, without whom this book wouldn't have been possible, in more ways than one.

I sincerely hope that you find this book useful. Please forward any comments or suggestions for improvement you may have to me. They will be greatly appreciated.

Zoher Z. Karu
zzkaru@alum.mit.edu

Table of Contents

Table of Contents

Table of Contents

Table of Contents

Table of Contents

CHAPTER 1 Introduction to Signals and Systems

Overview

This book is about the analysis of signals and how they interact with systems. We will focus on a special subclass of systems known as linear time-invariant systems. Signals and systems theory is the foundation of any engineer's knowledge. Just remember, everything is a signal, and all the world's a system.

1.1 Philosophy

Practically everything around us can be thought of as either a signal or a system. This book is about the study of signals and their interaction with systems. Although traditionally found in Electrical Engineering curricula, the subject of signals and systems is the foundation of all advanced engineering analysis. The concepts presented here are regularly used in a wide variety of fields ranging from fluid mechanics to economics.

This material presents an entirely new way of looking at the world around you – the frequency domain. This new concept provides a powerful alternative approach to traditional analytical methods. While learning about signals and systems, you will be presented with a variety of tools and techniques. Two concepts that may initially seem unrelated are likely to be deeply intertwined. Looking for these interconnections will help solidify the subject matter in your mind and illustrate the far-reaching significance of this important field.

1.2 What is a Signal?

A *signal* is an abstraction of any measurable quantity that is a function of one or more independent variables such as time or space. If that sounds vague, it's because signals are everywhere! Voltages and currents are examples of electrical signals; the sound coming into your ears is a mechanical signal; the page of this book is a two-dimensional light signal on your retina; the population of Nigeria is yet another type of signal.

There are two broad classes of signals: continuous time (CT) and discrete time (DT). A continuous-time signal is one that is present for all instants in time or space, such as an oscillating voltage signal or a photograph from a camera. A discrete-time signal is only present at discrete points in time or space. For example, the daily closing stock market average is a signal that changes only at discrete points in time (at the close of each day). A computer image composed of pixels is also a discrete time signal. Often, discrete-time signals are sampled versions of continuous-time signals, as is the case for the music recorded on compact discs or a photograph scanned into a computer.

1.3 What is a System?

In general, a *system* is an abstraction of anything that takes an input signal, operates on it, and produces an output signal. In other words, a system establishes a relationship between its input and its output. An example of a system is an automobile, where the input might be the position of the accelerator and the output the speed of the car. Another example is a camera, where the input is the light entering the lens and the output is a photograph. Systems that operate on continuous-time signals are known as CT systems, and systems that operate on discrete-time signals are known as DT systems.

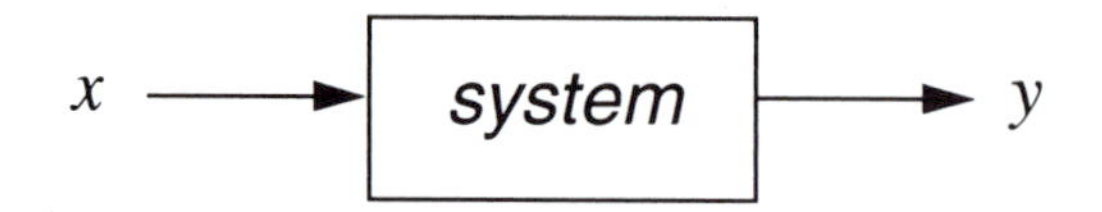

A special subclass of systems is both *linear* and *time-invariant*, known as LTI systems for short. In this book we will focus entirely on LTI systems. A system is considered to be linear if the following condition is met:

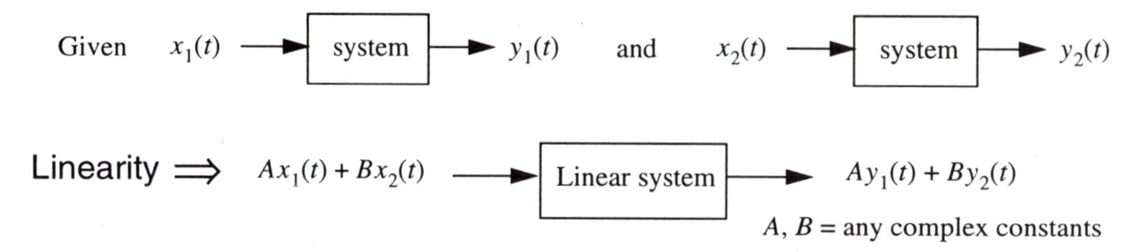

Also, for a linear system any linear operation on its input signal produces an output signal modified by the same linear operation. For example, a system that takes the absolute value of its input is not linear. An input of "2" produces an output of "2". However, an input of "-2" (multiply the original input by -1) does not produce an output of "-2". Examples of linear operations include scaling, integration, and differentiation.

A time-invariant system is one that responds the same no matter at what time the input is presented. If input "*x*" produces output "*y*", then the input "*x*" presented 5 days later will produce the same output "*y*" exactly 5 days later. In other words, delaying the input produces the corresponding delay in the output signal. Mathematically, this can be characterized as:

Given $x(t)$ → system → $y(t)$ TI ⇒ $x(t - \tau)$ → time-invariant system → $y(t - \tau)$

τ = time delay

Only a small percentage of the systems in the world are truly LTI. For example, a 5° deflection in the aileron of an airplane wing might produce a lift force of 1000N; will a 10° deflection produce a lift force of 2000N? Highly unlikely. So, then why are there literally hundreds of books on LTI system theory? Because nonlinear systems can be approximated as being linear within a small enough input range. The tools associated with LTI system analysis offer great insight into system behavior. They can fully characterize linear circuit elements such as ideal resistors, capacitors, and inductors or even ideal mechanical system elements like dashpots and springs. More complex nonlinear systems like the airplane aileron can still be analyzed using LTI techniques by first linearizing the system characteristics around an operating point; when the model breaks down, just move the operating point and linearize again. Other practical applications include the compact disc player, which is a prime example of signal processing and LTI system theory; by the end of this book you will have the tools to analyze how it works.

Now would be a good time to read through the Appendix of this book if you need a review of introductory circuit theory as well as some basic mathematical concepts.

CHAPTER 2 Continuous-Time Systems

Overview

Continuous-time systems are described by differential equations. Although traditionally decomposed into the particular and homogenous parts, for LTI system analysis it is more intuitive to break up the solution into what are known as the Zero State Response and Zero Input Response. The chapter also describes integrator-adder-gain block diagrams and state-space representations, two other methods for representing continuous-time systems.

2.1 A Differential Equation is a System

A system is anything that establishes a relationship between its input and its output. One way to describe a continuous-time system is through a differential equation. In this book, we will restrict ourselves to LTI systems, which correspond to linear, constant-coefficient differential equations such as:

$$\ddot{y} + 3\dot{y} + 2y = 2\dot{x} - x$$

In most cases, the input and output are both functions of time.

2.2 Finding the Solution

The complete solution to a differential equation consists of the sum of two parts: (1) the particular solution, which is any function that satisfies the differential equation, and (2) the homogenous solution, which is the solution of the differential equation with all input terms set to zero. Any unknown constants in the complete solution are usually found by substituting in initial conditions that specify the output and its derivatives at t=0.

The problem with the above solution process is that if the initial conditions are not zero ($y(0) \neq 0$), then the solution to the system is not linear. Recall that a linear system should produce zero output if the input is zero, which will not be the case if there is a nonzero initial condition. The tools that we will develop in this book are primarily for the analysis of linear systems. So, to get around this problem, we will separate the solution of any system into two parts: the Zero State Response (ZSR) and the Zero Input Response (ZIR). The system with all initial conditions (states) set to zero is said to be "at rest." It is a linear system, and its solution given an input signal is the ZSR. Meanwhile, the ZIR is the output of the system with the input set to zero; it is the response to the initial conditions only. The complete solution is then found by adding together the ZSR and ZIR. More of a discussion on solving differential equations for systems can be found in Section 3.5.1 of *Signals and Systems* by Oppenheim et al (1983).

Total Solution = ZSR + ZIR

Although they are closely related to the ZSR and ZIR, it is probably best to forget about particular and homogenous solutions and start thinking in terms of these new concepts. Finding the ZSR and ZIR is the preferred method of system analysis since they provide a more intuitive understanding of system behavior. There will be a more complete discussion of the ZSR and ZIR in Chapter 5.

2.3 Integrator-Adder-Gain Block Diagrams

Continuous-time systems are sometimes described graphically in what are known as integrator-adder-gain block diagrams. After manipulating the original differential equation into the correct form by repeatedly integrating both sides of the equation to remove all derivative terms, it is easy to draw the correct block diagram by simply plugging in the appropriate coefficients into the canonical form shown below. For its derivation, see Problem 1.6 on page 34 of *Circuits, Signals, and Systems* by Siebert (1986).

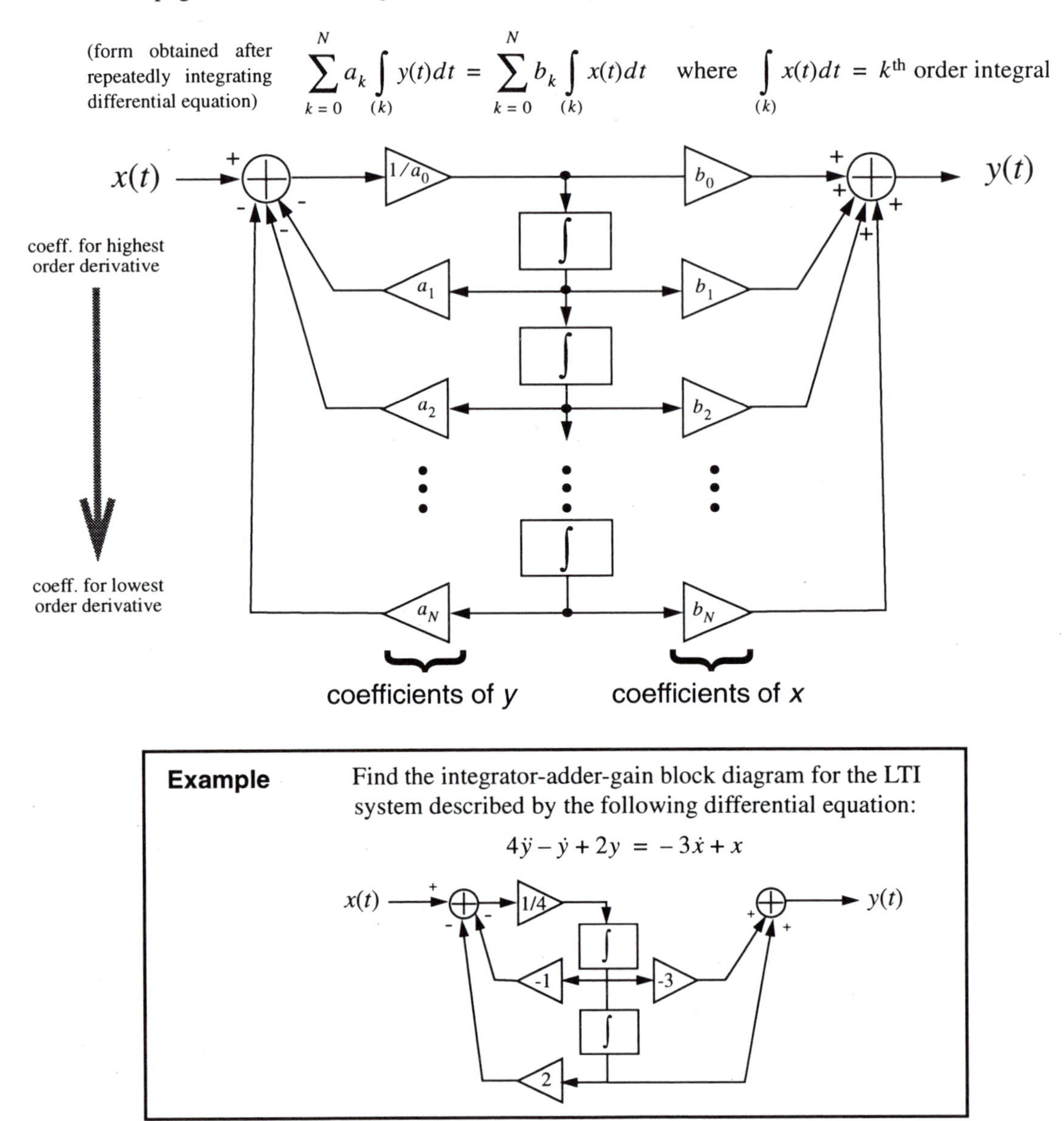

2.4 State Equations

A state-space representation is yet another method of describing a continuous-time system. A "state" is any internal system variable whose next value depends on its current value; in other words, it contains information about how the system evolves in time. For example, in an RLC circuit, capacitor voltages and inductor currents are state variables. The higher order differential equation that describes the overall system can be written as a series of first order differential equations relating the inputs, outputs, and these internal states. As an example, let's analyze the following circuit:

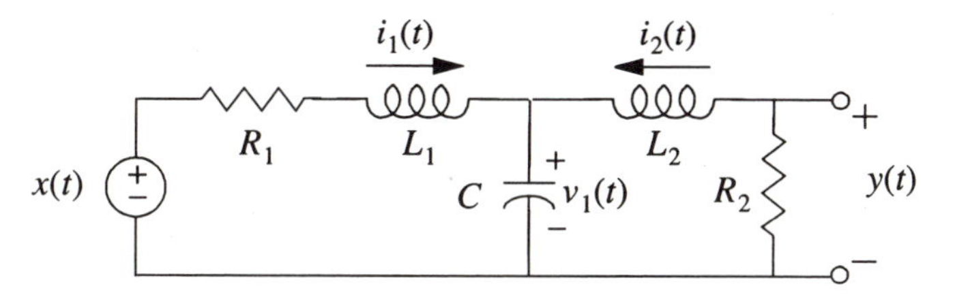

Identify the three state variables $i_1(t)$, $i_2(t)$, and $v_1(t)$.

They are related through the following three first order differential equations:

$$L_1\frac{di_1(t)}{dt} = -R_1 i_1(t) - v_1(t) + x(t) \qquad \text{(KVL)}$$

$$L_2\frac{di_2(t)}{dt} = -R_2 i_2(t) - v_1(t) \qquad \text{(KVL)}$$

$$C\frac{dv_1(t)}{dt} = i_1(t) + i_2(t) \qquad \text{(KCL)}$$

The series of state equations can be compactly written in matrix form, known as the *state-space representation.*

$$\frac{d}{dt}\begin{bmatrix} i_1(t) \\ i_2(t) \\ v_1(t) \end{bmatrix} = \begin{bmatrix} -R_1/L_1 & 0 & -1/L_1 \\ 0 & -R_2/L_2 & -1/L_2 \\ 1/C & 1/C & 0 \end{bmatrix}\begin{bmatrix} i_1(t) \\ i_2(t) \\ v_1(t) \end{bmatrix} + \begin{bmatrix} 1/L_1 \\ 0 \\ 0 \end{bmatrix} x(t) \qquad y(t) = \begin{bmatrix} 0 & -R_2 & 0 \end{bmatrix}\begin{bmatrix} i_1(t) \\ i_2(t) \\ v_1(t) \end{bmatrix}$$

(3 states, 1 input, 1 output)

In general, the form is:

State-Space Representation

$$\dot{x} = Ax + Bu$$

$$y = Cx + D$$

$x = N$ states $(N \times 1)$

$u = M$ inputs $(M \times 1)$

$y = K$ outputs $(K \times 1)$

A = system matrix $(N \times N)$

B, C = coefficients $(N \times M$ and $K \times N)$

D = constants $(K \times 1)$

Writing systems in state-space representation has many advantages. The matrix form makes it very easy to represent and solve the system on a computer. Also, there are many system properties such as stability that can be determined by examining "A", the system matrix. The true power of a state-space representation is best realized when analyzing multiple-input, multiple-output systems, where writing down every equation would otherwise become unmanageable.

CHAPTER 3 The Frequency Domain

Overview

The frequency domain is probably the most important concept in signal and system theory. Systems often become much easier to analyze by expressing the input as a function of frequency rather than time. The relationship between the time and frequency domains is the great unifying concept of this subject. This chapter introduces complex frequency and the s-plane and then explains their role in continuous-time system analysis through the notions of eigenfunctions and impedance. The chapter concludes with some examples of how quickly traditional steady-state problems can be solved.

3.1 Why a New Domain?

We have been talking about systems strictly in the time domain, where the inputs and outputs were represented as functions of time. However, as we shall soon see, it is often much easier to analyze signals and systems when they are represented in the *frequency domain*. The entire subject of signals and systems consists primarily of the following concepts: (1) writing signals as functions of frequency; (2) looking at how systems respond to inputs of different frequencies; (3) developing tools for switching between time-domain and frequency-domain representations; and (4) learning how to determine which domain is best suited for a particular problem.

3.2 Complex Frequency

You have most likely only heard of the word "frequency" applied to the number of cycles per second of a periodic signal. Well, it's time to broaden your horizons. For complete generality, we will allow this "frequency" to be a complex number. Huh? Complex frequency? What are you talking about? Just loosen your definition of the word "frequency" and keep reading. We will hereby declare that any function written in the form Ke^{st} (where K and s are in general complex constants) is characterized by the complex frequency "s", which can be expanded as $s = \sigma + j\omega$. The following table provides examples of describing signals using this new frequency "s".

Type of Signal	*Example*	*Frequency*
DC	$x(t) = 5$	$s = 0$
Exponential	$x(t) = e^{-3t}$	$s = -3$

Type of Signal	*Example*	*Frequency*
Sinusoidal	$x(t) = \sin 50t = \frac{1}{2j}(e^{50jt} - e^{-50jt})$	$s = \pm 50j$
Exponential Sinusoid	$x(t) = e^{2t}\cos 100t = \frac{1}{2}(e^{(2+100j)t} + e^{(2-100j)t})$	$s = 2 \pm 100j$

To obtain the last two entries, recall that $e^{j\theta} = \cos\theta + j\sin\theta$. The table shows how either a single real value or pair of complex conjugate values of "s" is able to completely characterize a wide variety of types of functions. It is now possible to visualize what is commonly called "the s-plane." In the plot below, the shape of the time-domain function is sketched in the region of the complex frequency plane to which it corresponds:

The S-Plane

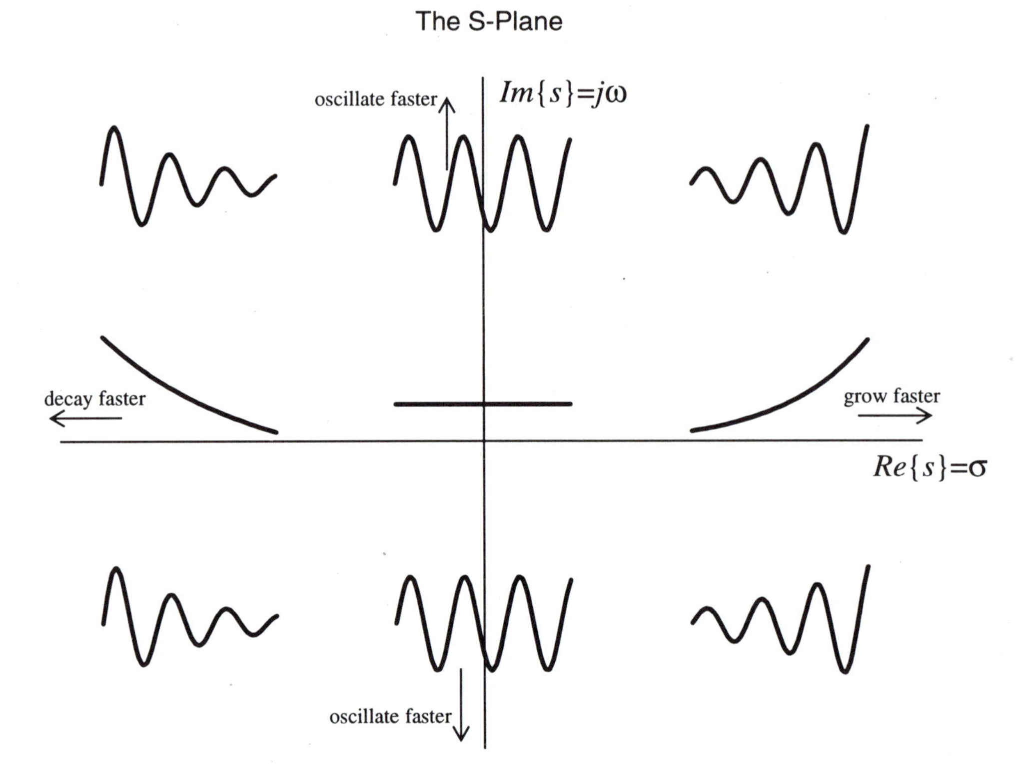

A sinusoidal or exponential-sinusoidal function can also be expressed as the real part of a single complex exponential as shown below:

$$A\cos(\omega t + \theta) = ARe[e^{j\omega t}e^{j\theta}] = Re[Be^{j\omega t}] \quad (B \text{ is complex})$$

So, using the principles of linearity and superposition, we can find a system's response to a sinusoidal input by first finding its response to the corresponding complex exponential input, and then just taking the real part of the answer. This may initially seem like a lot more work, but this procedure will be much easier in the long run since functions of the form e^{st} play a special role in CT systems analysis.

3.3 Eigenfunctions

Functions of the form e^{st} are known as *eigenfunctions* of continuous-time LTI systems. This means that when such a function is an input to a CT LTI system, the output is a function of the exact same complex frequency as the input, except it is multiplied by a scaling factor.

Output has same frequency as input!

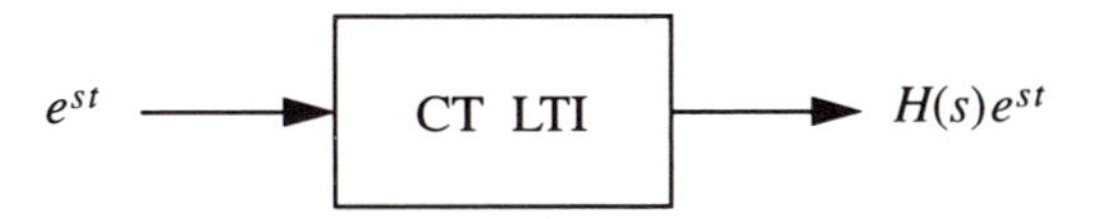

The coefficient $H(s)$ is in general a complex number, and its value depends on the frequency "s" of the input signal. Multiplying a function by a complex number alters its magnitude and its phase. For example, if $\sin(\omega t)$ goes in, then $A\sin(\omega t + \theta)$ comes out.

It is actually quite easy to prove that forms of e^{st} are indeed eigenfunctions, but only after developing the concepts in the next several chapters; the proof will have to wait until Section 10.4. Please note that functions of the form $e^{st}u(t)$ (where $u(t)$ is the unit step function as defined in Section 9.4) are <u>not</u> eigenfunctions of CT LTI systems. This is a common error so watch out for it!

Remember, for eigenfunction inputs to CT LTI systems, the same complex frequency must appear at the output. Use that fact to determine whether or not the following systems could be LTI:

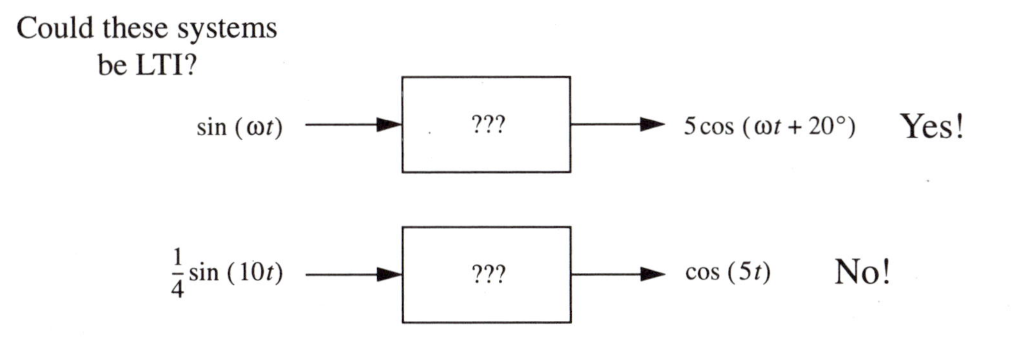

3.4 An Example Problem

Let's use the concepts of complex frequency and eigenfunctions to help us solve the following problem:

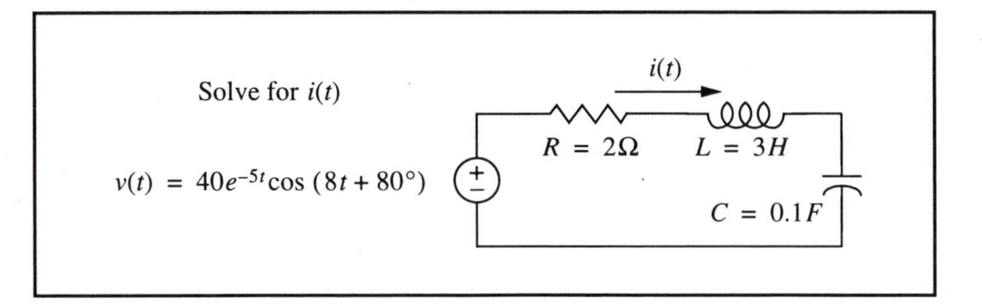

Solution

First, express the input in the form of a complex exponential (variables with a ^ mean they are complex). For the mathematically strict, use radians, not degrees in the phase.

$$v(t) = Re\{40e^{j80^\circ}e^{(-5+8j)t}\}$$

$$\hat{v}(t) = 40e^{j80^\circ}e^{(-5+8j)t}$$

$$= \hat{V}e^{s_0 t} \quad (\hat{V} = 40e^{j80^\circ},\ s_0 = -5+8j)$$

Write the KVL equation.

$$\hat{v}(t) = R\hat{i}(t) + L\frac{d\hat{i}(t)}{dt} + \frac{1}{C}\int_{-\infty}^{t}\hat{i}(\tau)d\tau$$

By eigenfunction theory, we know that the output must be a signal of the same s-plane frequency with a new complex scaling factor.

$$\hat{i}(t) = \hat{I}e^{s_0 t}$$

Substitute in the complex exponential forms of the input and output.

$$\hat{V}e^{s_0 t} = R\hat{I}e^{s_0 t} + Ls_0\hat{I}e^{s_0 t} + \frac{1}{Cs_0}\hat{I}e^{s_0 t}$$

Substitute values for R, L, C and divide through by the common factor $e^{s_0 t}$.

$$\hat{V} = 2\hat{I} + 3s_0\hat{I} + \frac{10}{s_0}\hat{I}$$

Substitute in the value for s_0 and solve for $\hat{I}$.

$$\hat{I} = \frac{\hat{V}}{2+3s_0+10/s_0} = \frac{40e^{j80^\circ}}{2+3(-5+8j)+10/(-5+8j)}$$

Complex number manipulation...

$$\hat{I} = \frac{40e^{j80^\circ}}{-13+24j+(-50-80j)/89} = \frac{40e^{j80^\circ}}{-13.56+23.10j}$$

$$\hat{I} = \frac{40e^{j80^\circ}}{26.79e^{j120.41^\circ}} = (40/26.79)e^{j(80^\circ-120.41^\circ)} = 1.49e^{-j40.41^\circ}$$

Finally the answer!

$$\boxed{i(t) = Re\{\hat{I}e^{s_0 t}\} = 1.49e^{-5t}\cos(8t-40.41^\circ)}$$

Note that we were able to find the solution without the traditional method of solving the differential equation (assuming a form of the solution, etc.). Although this new procedure may still seem a bit complicated, it will become much easier after introducing the concepts in the next few sections. We will return to this type of problem in Section 3.6 to illustrate just how quickly they can be solved. If you are not comfortable with complex number manipulation, please read the Appendix of this book now.

3.5 Impedance

In the previous section we found the current flowing through a circuit given the applied voltage. Using Ohm's Law of $V = IR$, we could have found the equivalent "resistance" of the circuit by dividing V by I. Since the term resistance officially applies only to resistors, for general circuits containing R's, L's, and C's we will call this quantity *impedance* designated by the symbol Z.

> Impedance Form of Ohm's Law
>
> $$V = IZ$$

The impedance is in general a complex number and its value will depend on the complex frequency of the input signal. For example, let's study a single capacitor and identify its impedance:

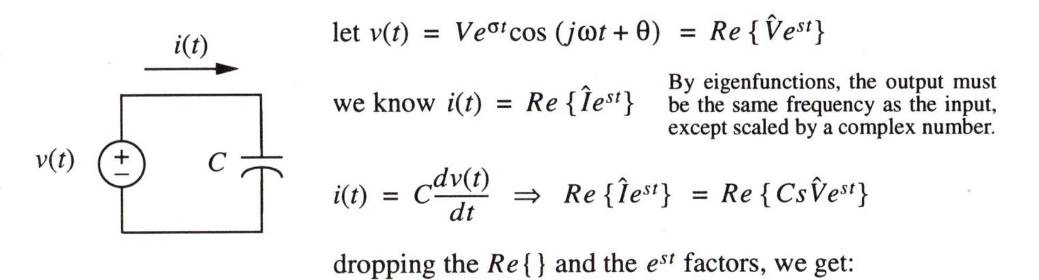

let $v(t) = Ve^{\sigma t}\cos(j\omega t + \theta) = Re\{\hat{V}e^{st}\}$

we know $i(t) = Re\{\hat{I}e^{st}\}$ By eigenfunctions, the output must be the same frequency as the input, except scaled by a complex number.

$$i(t) = C\frac{dv(t)}{dt} \Rightarrow Re\{\hat{I}e^{st}\} = Re\{Cs\hat{V}e^{st}\}$$

dropping the $Re\{\}$ and the e^{st} factors, we get:

$$Z(s) = \frac{\hat{V}}{\hat{I}} = \frac{1}{sC}$$

The impedance of a capacitor is $1/(sC)$. Recall from basic circuit theory that a capacitor tends to block low frequency signals (like DC) and easily pass high frequency signals. Plug in s=small and s=big into the impedance expression and you should now understand why. Impedances for other circuit elements can be derived in a similar fashion; the results are summarized in the table below:

Element	***Impedance***
Resistor (R)	R
Inductor (L)	sL
Capacitor (C)	$\frac{1}{sC}$

Remember, the impedance of a circuit is not a constant; it depends on the frequency of the input signal.

3.6 Steady-State Analysis

Inputs of the form e^{st} are steady-state inputs, i.e. they exist for all time $-\infty < t < \infty$. Using the concept of eigen-functions, we know that the steady-state output must be of the same frequency as the input, but with a possible change in magnitude and phase. That insight allows us to solve steady-state problems quite easily.

Given input = $Re\{\hat{V}_{in}e^{st}\}$, we know that output = $Re\{\hat{V}_{out}e^{st}\}$ where $\hat{V}_{out} = H(s)\cdot\hat{V}_{in}$

(complex # — complex #: to find the product, multiply the magnitudes and add the phases)

Example:

Input = $A\cos(\omega t + \theta) \Rightarrow$ Output = $A|H(j\omega)|\cos(\omega t + \theta + \angle H(j\omega))$

Now, solving a steady-state problem is merely reduced to finding the magnitude and phase of $H(s)$ evaluated at the proper complex frequency. The concept of impedance will make finding $H(s)$ much easier. Observe:

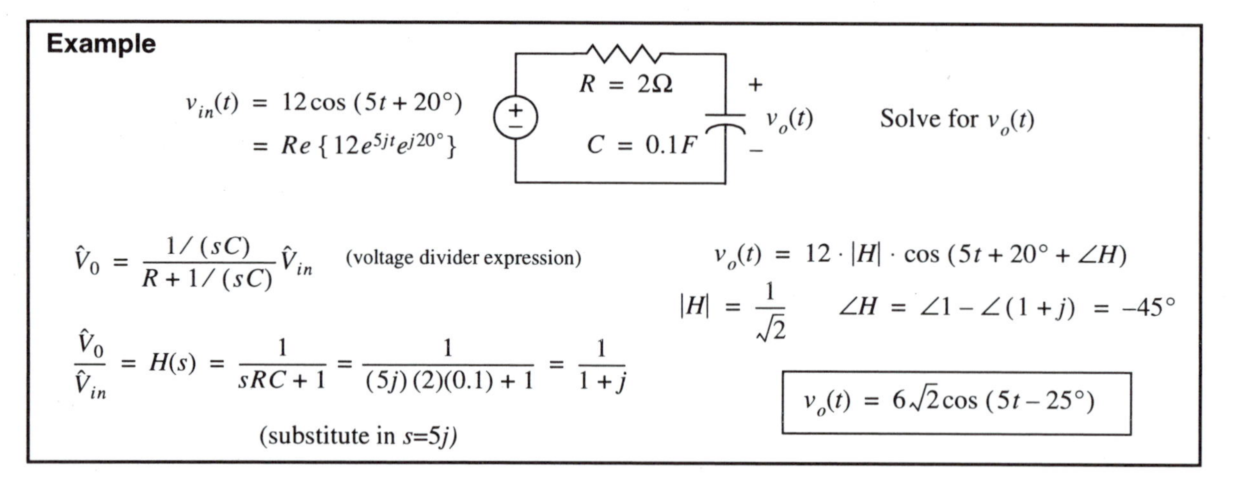

Example

$$v_{in}(t) = 12\cos(5t + 20°) = Re\{12e^{5jt}e^{j20°}\}$$

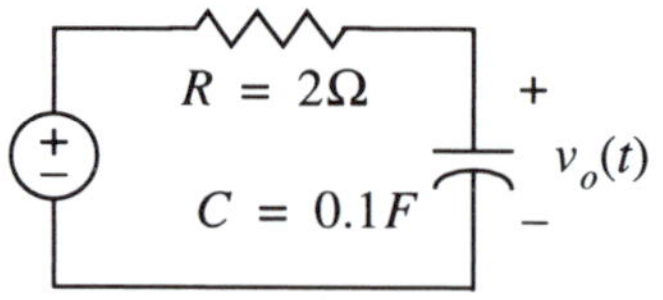

Solve for $v_o(t)$

$$\hat{V}_0 = \frac{1/(sC)}{R + 1/(sC)}\hat{V}_{in} \quad \text{(voltage divider expression)}$$

$$\frac{\hat{V}_0}{\hat{V}_{in}} = H(s) = \frac{1}{sRC+1} = \frac{1}{(5j)(2)(0.1)+1} = \frac{1}{1+j}$$

(substitute in $s=5j$)

$$v_o(t) = 12\cdot|H|\cdot\cos(5t + 20° + \angle H)$$

$$|H| = \frac{1}{\sqrt{2}} \qquad \angle H = \angle 1 - \angle(1+j) = -45°$$

$$\boxed{v_o(t) = 6\sqrt{2}\cos(5t - 25°)}$$

Example

$$v_{in}(t) = 12\cos(5t + 20°) = Re\{12e^{5jt}e^{j20°}\}$$

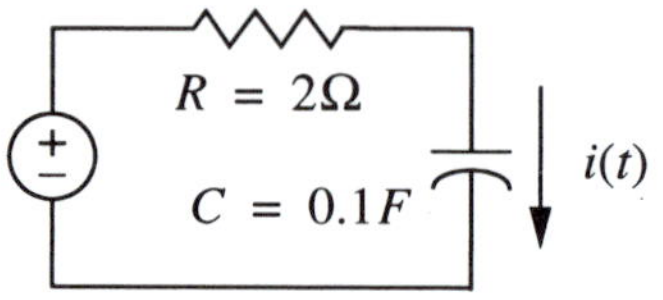

Solve for $i(t)$

$$\hat{V}_{in} = \left(R + \frac{1}{sC}\right)\hat{I} \quad \text{(Ohm's Law)}$$

$$\frac{\hat{I}}{\hat{V}_{in}} = H(s) = \frac{1}{R + \frac{1}{sC}} = \frac{1}{2 + \frac{1}{(5j)(0.1)}} = \frac{1}{2-2j}$$

(substitute in $s=5j$)

$$i(t) = 12\cdot|H|\cdot\cos(5t + 20° + \angle H)$$

$$|H| = 1/(2\sqrt{2}) \qquad \angle H = 45°$$

$$\boxed{i(t) = 3\sqrt{2}\cos(5t + 65°)}$$

Note: this is the same answer as $C\frac{dv_0(t)}{dt}$

Go back and try the example in Section 3.4 again. This time you should be able to skip several of the steps and arrive at the answer much more quickly.

CHAPTER 4 The Laplace Transform

Overview

The Laplace transform is the most important tool in the analysis of continuous-time systems. It is used to decompose a signal into the sum of complex exponentials, each of which is an eigenfunction of an LTI system. Since the output of a system for an eigenfunction input is easily determined, the Laplace transform can be used to determine the output for virtually any input, thus greatly simplifying the analysis of CT systems. This chapter introduces the forward and inverse transform and talks about many of its mathematical properties.

4.1 What is the Laplace Transform?

We have seen how to analyze systems that have complex exponential inputs, but obviously there are other types of signals. What should we do then? The Laplace transform is a mathematical operation that can express practically any continuous-time signal as the sum of complex exponentials of the form Ke^{st}. Since we know the response to each e^{st} eigenfunction using the principles in Chapter 3, we can reconstruct the output for virtually any input by using the principles of linearity and superposition inherent in the Laplace transform. As we shall soon see, the Laplace transform greatly simplifies the analysis of continuous-time systems.

The formulas for the forward and inverse transform are shown below. Looking carefully at the formula for the inverse transform, we see that it basically says that a signal $x(t)$ can be expressed as the sum (integral) of an infinite number of appropriately scaled complex exponentials of the form $X(s)e^{st}$.

Bilateral Laplace Transform

$$X(s) = \int_{-\infty}^{\infty} x(t)e^{-st}dt$$

Inverse Laplace Transform

$$x(t) = \frac{1}{2\pi j}\int_{\sigma-j\infty}^{\sigma+j\infty} X(s)e^{st}ds$$

For the inverse transform, the contour of integration is a straight line parallel to the $j\omega$-axis for any value of σ in the transform's region of convergence (defined in Section 4.3). But don't panic; there's an easier way to take the inverse transform, as shown in Section 4.6. We will spend the rest of this chapter describing the mathematical properties of the Laplace transform and some example forward and inverse pairs. In Chapter 5, we will begin to illustrate the role of the Laplace transform in the analysis of CT systems.

4.2 Unilateral vs. Bilateral

In this book we will use what is known as the *bilateral* Laplace transform. This is the formula given in Section 4.1. Another form of the Laplace transform is known as the *unilateral* form, where the limits of integration are only from zero to infinity. Most people concerned with the "real world" (e.g. control system designers) almost always use the unilateral formula since to them there is no such thing as negative time. But the theorists and the non-real time signal processors usually use the bilateral form primarily because of its appealing mathematical properties. In the bilateral world, signals that start at time=0 are generally written in combination with the unit step function, such as $x(t)u(t)$. Forgetting to add the $u(t)$ is a common error, so watch out for it. Also, remember that an input of the form e^{st} is composed of a single complex frequency, whereas something like $e^{st}u(t)$ is composed of many different complex frequencies. The only significant difference between the unilateral and bilateral transforms is the way they deal with initial conditions, which is evident in the integration and differentiation properties in Section 4.4.

4.3 Region of Convergence

The region of convergence (ROC) of a Laplace transform is defined as the set of values of s for which the Laplace transform integral can be evaluated (i.e. it doesn't blow up).

Example If $x(t) = e^{2t}u(t)$, find $X(s)$ and its ROC.

$$X(s) = \int_{-\infty}^{\infty} e^{2t}e^{-st}u(t)dt = \int_{0}^{\infty} e^{(2-s)t}dt = \frac{1}{2-s}e^{(2-s)t}\Big|_0^{\infty} = \frac{1}{s-2}$$

s-plane

ROC = shaded region

only possible if (2-$Re\{s\}$) in exponent < 0

$\Rightarrow$ ROC is $Re\{s\}>2$

There are several rules that can be used to quickly determine the ROC without performing any integration; all that is needed is the type of signal (see chart below) and the location of the poles of the transform. A pole is defined as a value of s that causes the denominator of the transform to become zero. The concept of a pole will be more clearly defined and explained in Chapter 5.

Types of Signals

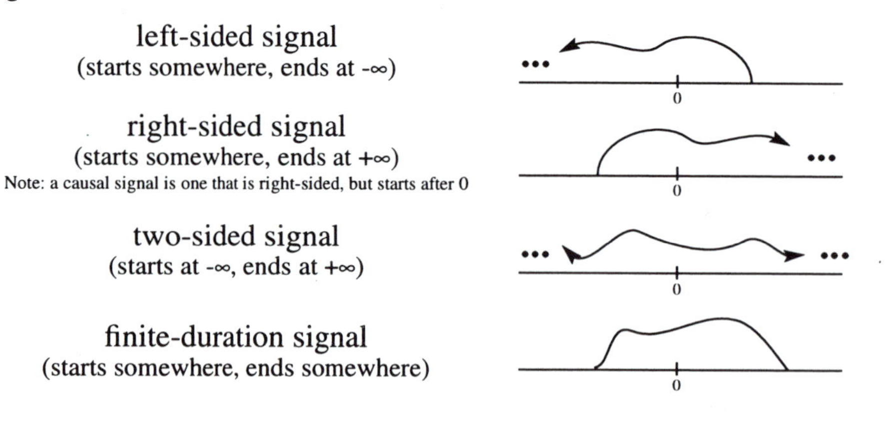

Rules for the ROC

- The ROC is always a region of the s-plane to the left or right of a vertical line, or a strip between two vertical lines.
- The ROC never contains any poles.
- If $x(t)$ is right-sided, then the ROC is right-sided, i.e. $Re\{s\} > a$, where a is the Re\{rightmost pole\}.
- If $x(t)$ is left-sided, then the ROC is left-sided, i.e. $Re\{s\} < a$, where a is the Re\{leftmost pole\}.
- If $x(t)$ is two-sided or the sum of a left and right sided signal, the ROC is either a strip ($a < Re\{s\} < b$), or else the individual ROC's will not overlap, producing the null set.
- If $x(t)$ is of finite duration, then the ROC is the entire s-plane.

4.4 Bilateral Laplace Transform Properties

Property	$x(t)$	$X(s)$	*New ROC*
Linearity	$ax_1(t) + bx_2(t)$	$aX_1(s) + bX_2(s)$	ROC $\supseteq$ ROC(x_1) $\cap$ ROC(x_2)
Time Shift	$x(t - t_o)$	$e^{-st_0}X(s)$	ROC(x)
Exponential Multiply	$e^{-\alpha t}x(t)$	$X(s + \alpha)$	shift ROC to left by α
Times t	$tx(t)$	$-\frac{dX(s)}{ds}$	ROC(x)
Time Scaling	$x(at)$	$\frac{1}{\|a\|}X(s/a)$	Scaled ROC (s in new ROC if s/a in old ROC)
Integration	$\int_{-\infty}^{t} x(\tau)d\tau$	$\frac{1}{s}X(s)$	ROC $\supseteq$ ROC(x) $\cap$ $Re\{s\}>0$
Differentiation	$\frac{dx(t)}{dt}$	$sX(s)$	ROC $\supseteq$ ROC(x)

Differences for the Unilateral Transform

Integration	$\int_0^t x(\tau)d\tau$	$\frac{1}{s}X(s)$
Differentiation	$\frac{dx(t)}{dt}$	$sX(s) - x(0)$

4.5 Table of Transform Pairs

The following table lists some common Laplace transform pairs; transforms for other common signals can be derived using the properties in Section 4.4

$x(t)$	$X(s)$	*ROC*
$\delta(t)$	1	All values of s
$u(t)$	$\frac{1}{s}$	$Re\{s\} > 0$
$e^{-\alpha t}u(t)$	$\frac{1}{s+\alpha}$	$Re\{s\} > -\alpha$
$-e^{-\alpha t}u(-t)$	$\frac{1}{s+\alpha}$	$Re\{s\} < -\alpha$
$t^n u(t)$	$\frac{n!}{s^{n+1}}$	$Re\{s\} > 0$
$(\sin\omega_0 t)\, u(t)$	$\frac{\omega_0}{s^2+\omega_0^2}$	$Re\{s\} > 0$
$(\cos\omega_0 t)\, u(t)$	$\frac{s}{s^2+\omega_0^2}$	$Re\{s\} > 0$

4.6 Inverse Laplace Transform

The general formula for recovering $x(t)$ from $X(s)$ is the complex line integral:

$$x(t) = \frac{1}{2\pi j}\int_{\sigma - j\infty}^{\sigma + j\infty} X(s)e^{st}ds$$

DO NOT USE THIS FORMULA!!!

Rather, to perform the inverse Laplace transform, we will merely manipulate the given expression until we see patterns we recognize from the Laplace transform table. This is basically a heuristic scheme and is one that will become more obvious with practice.

Note that knowing the ROC is critical to performing an inverse Laplace transform. For example:

$$X(s) = \frac{1}{s+3} \quad ? \quad \begin{cases} x(t) = e^{-3t}u(t) & \text{ROC: } Re\{s\} > -3 \\ x(t) = -e^{-3t}u(-t) & \text{ROC: } Re\{s\} < -3 \end{cases}$$

More complex transforms can be evaluated by first splitting the transform into simpler expressions using the process of partial fraction expansion (PFE). See the Appendix for a short review of this important mathematical technique. The following example illustrates a simple PFE and the importance of utilizing information about the region of convergence.

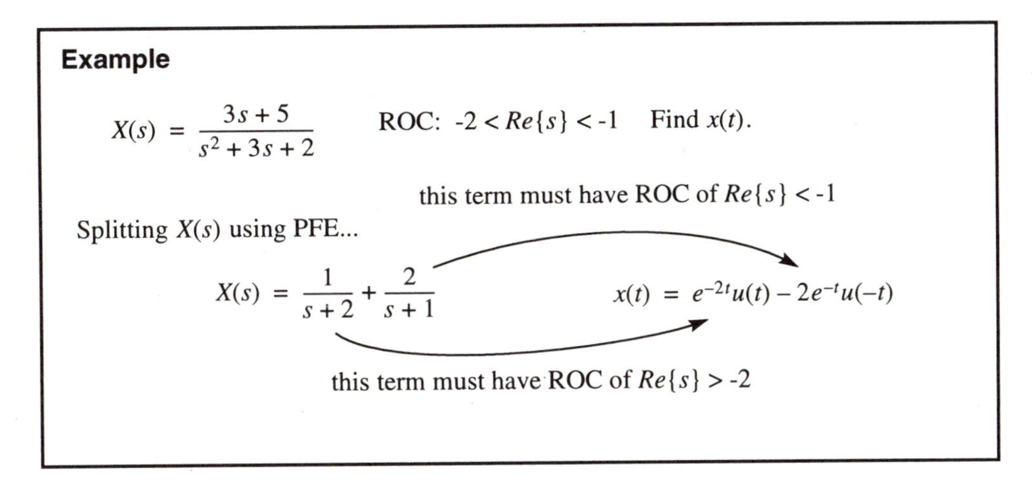

Example

$$X(s) = \frac{3s+5}{s^2+3s+2} \qquad \text{ROC: } -2 < Re\{s\} < -1 \qquad \text{Find } x(t).$$

Splitting $X(s)$ using PFE...

$$X(s) = \frac{1}{s+2} + \frac{2}{s+1} \qquad x(t) = e^{-2t}u(t) - 2e^{-t}u(-t)$$

Sometimes however, factoring the denominator is not always the best thing to do. In the case of a second order polynomial with complex roots, completing the square is generally the best procedure. Observe:

Example

$$X(s) = \frac{1}{s^2+4s+40} \qquad \text{ROC: } Re\{s\} > -2 \qquad \text{Find } x(t).$$

$$X(s) = \frac{1}{s^2+4s+4+36} = \frac{\left(\frac{1}{6}\right)6}{(s+2)^2+6^2} \qquad \Rightarrow x(t) = \frac{1}{6}e^{-2t}\sin(6t)\,u(t)$$

4.7 Initial and Final Value Theorems

Given only a rational transform $X(s)$, the initial and final value theorems allow us to determine the values of $x(t=0^+)$ and $x(t=\infty)$ without having to go back to the time domain through an inverse transform. Be sure to keep in mind the conditions under which these theorems are valid.

Initial Value Theorem	$x(t=0^+) = \lim_{s\to\infty} sX(s)$	$x(t) = 0,\ t < 0$ and no impulses or higher order singularities at origin
Final Value Theorem	$x(t=\infty) = \lim_{s\to 0} sX(s)$	$X(s)$ has no poles in $Re\{s\} \geq 0$ (one pole at $s=0$ ok); ROC is right-sided

CHAPTER 5 CT Systems Analysis

Overview

This chapter presents the underlying concepts and methods in the analysis of continuous-time LTI systems. A system's entire characteristics can be captured in a single function $H(s)$, appropriately known as the system function. This chapter defines the notions of poles and zeros and illustrates how they govern system behavior. Example problems illustrate the use of the Laplace transform in determining the ZIR and ZSR, the two components of the solution to any CT LTI system. Finally, the impulse response is briefly introduced in an effort to begin an appreciation for how nicely everything really does fit together.

5.1 The System Function

Recall from the inverse Laplace transform formula that a signal $x(t)$ can be represented as the infinite sum (integral) of appropriately scaled complex exponentials. We will now illustrate why this makes the Laplace transform a powerful tool for the analysis of CT systems. Not being mathematically strict, let's rewrite the inverse Laplace transform integral as a Riemann sum:

$$x(t) = \frac{1}{2\pi j}\int_{\sigma-j\infty}^{\sigma+j\infty} X(s)e^{st}ds \approx \frac{1}{2\pi j}\{\ldots + X(s_0)e^{s_0 t} + X(s_1)e^{s_1 t} + \ldots + X(s_\infty)e^{s_\infty t}\}\,\Delta s$$

Now recall that complex exponentials of the form e^{st} are eigenfunctions of a continuous-time LTI system:

$$X(s)e^{st} \longrightarrow \boxed{\text{CT LTI}} \longrightarrow X(s)H(s)e^{st}$$

Utilizing this property along with linearity allows us to relate $y(t)$, the output of an LTI system, to the input $x(t)$ as follows:

$$y(t) \approx \frac{1}{2\pi j}\{\ldots\ldots + H(s_0)X(s_0)e^{s_0 t} + H(s_1)X(s_1)e^{s_1 t} + \ldots\ldots + H(s_\infty)X(s_\infty)e^{s_\infty t}\}\,\Delta s$$

$$\boxed{y(t) = \frac{1}{2\pi j}\int_{\sigma-j\infty}^{\sigma+j\infty} H(s)X(s)e^{st}ds = \frac{1}{2\pi j}\int_{\sigma-j\infty}^{\sigma+j\infty} Y(s)e^{st}ds}$$

You may not have realized it, but we have just revealed two of the most fundamental concepts in linear system theory. The first is that $H(s)$ is a complete characterization of the system. That's why $H(s)$ is commonly referred to as *the system function* (sometimes called a transfer function). If we know $H(s)$ for all values of possible complex frequencies "s", then we can find the response to all possible complex exponential inputs. Since practically any CT signal can be broken down into the sum of complex exponentials by utilizing the Laplace transform, given $H(s)$ we should be able to find the output response for any possible input signal. The second fundamental concept we have discovered is that this output signal can be obtained through the simple relationship $Y(s) = H(s)X(s)$.

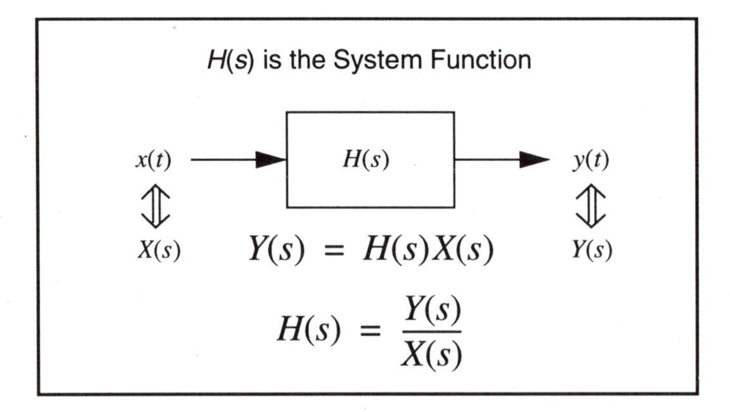

5.2 Poles and Zeros

The concept of poles and zeros is best introduced through an example. Consider the following system:

$v_i(t)$ 0.01F 10Ω $v_o(t)$

$$V_0(s) = \frac{R}{R + 1/(sC)} V_i(s) \qquad \text{(voltage divider)}$$

$$H(s) = \frac{V_o(s)}{V_i(s)} = \frac{R}{R + 1/(sC)} = \frac{s}{s + 1/(RC)} = \frac{s}{s + 10}$$

Remember, $s=\sigma+j\omega$. Let's see what happens when we plot $|H(s)|$ versus σ while letting $\omega=0$.

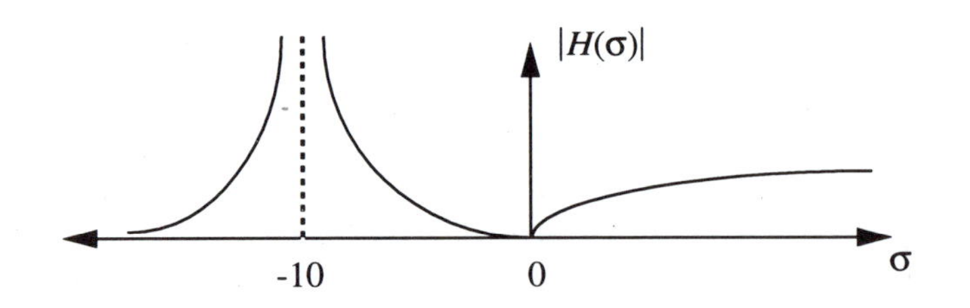

The frequency $s= -10$ is a *pole* of $H(s)$ since $H(s)$ evaluated at that point is infinitely high. In the plot shown above, it looks like there is a stick (pole) placed at $s= -10$ holding up the graph. Similarly, the frequency $s=0$ is known as a *zero* of $H(s)$ since $H(s)$ evaluates to zero at that point. Poles and zeros specify the set of complex frequencies for which the eigenfunction response is infinite or zero, respectively. For example, the above system's response to e^{-10t} (which is the same as the steady-state response to $e^{-10t}u(t)$) is infinite. Putting Ke^0 (a DC input) into the system will result in an output of zero, but that should have been obvious from looking at the circuit anyhow. Note that putting $Ku(t)$ into the system will initially produce a transient output, but the steady state (t=∞) output is still zero since $u(t)$ looks like a DC input after a long period of time.

A *pole-zero plot* shows the locations of the poles (drawn using X's) and zeros (drawn using O's) on the s-plane. The pole-zero plot for the previously described circuit is shown below:

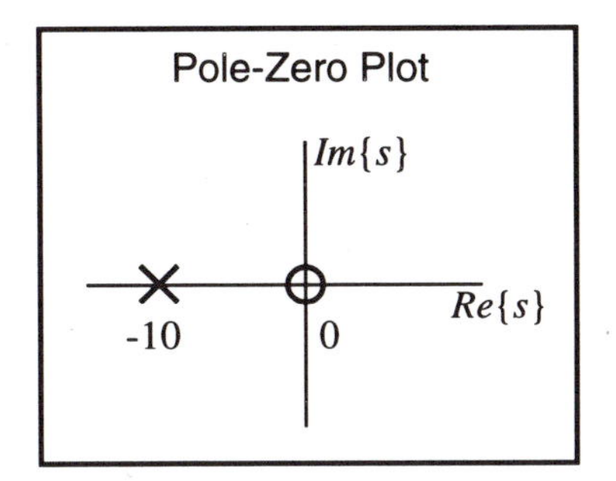

Note that poles and zeros can occur anywhere in the s-plane. Whenever the poles and zeros are the roots of polynomials with real coefficients, the poles and zeros off the real axis will always occur in complex conjugate pairs. Also note that there can be more than one pole or zero at the same location, but obviously a pole and a zero at the same place merely cancel each other out (numerator/denominator cancellation). Given only the pole-zero plot, it is possible to reconstruct $H(s)$, but only to within a constant scaling factor. For example:

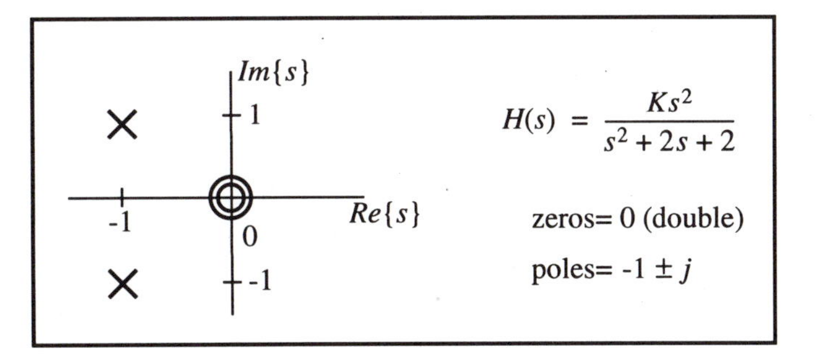

The number of poles in a system will always correspond to the number of independent state variables in the system, i.e. how many initial conditions must be specified. The number of poles is known as the *order* of the system. Note that the number of zeros does not affect system order.

Order of system = number of poles

5.3 Converting Differential Equations to System Functions

Earlier we said that a differential equation is one way of describing a system. We have seen that $H(s)$ is also a complete characterization of a system, so there must be an easy way of converting between these two forms. All it involves is taking the Laplace transform of both sides of the equation (using the derivative property of the transform where appropriate) and manipulating the result to find an expression for $Y(s)/X(s)$. Given an $H(s)$, the differential equation can be recovered by cross-multiplying and doing an inverse transform. For practice, try doing the following example in both directions.

Example

$$2\ddot{y} - 3\dot{y} + 5y = 10\dot{x} - 7x$$ original differential equation

$$2s^2Y(s) - 3sY(s) + 5Y(s) = 10sX(s) - 7X(s)$$ take the Laplace transform of both sides

$$H(s) = \frac{Y(s)}{X(s)} = \frac{10s - 7}{2s^2 - 3s + 5}$$ simplify and solve for $Y(s)/X(s)$

5.4 Zero Input Response

The Zero Input Response (ZIR) is the term given to the output of a system when there is no input signal present. This means that the ZIR corresponds to the manner in which any initial conditions present decay away.

10Ω 0.01F + $v_o(t)$ −

if $v_0(0) = 5V$, ZIR $= 5e^{-10t}u(t)$

How were we able to find the form of the ZIR for the above circuit? Recall that it is always the case that $Y(s) = H(s)X(s)$. If there is no input, this means that $X(s)=0$. Does that automatically mean that $Y(s)$ must be zero? No! A finite value of $Y(s)$ when $X(s)=0$ is possible only if $H(s)$ is infinitely large, meaning that the signal $y(t)$ contains a complex frequency that is a pole of $H(s)$. Thus, the ZIR can only contain frequencies that are poles of $H(s)$. These frequencies are referred to as the *natural frequencies* of the system since the system is naturally tuned to oscillate or decay at those frequencies in the absence of an input signal. The equation obtained by setting the denominator of the system function equal to zero is known as the *characteristic equation* since its roots (the poles of the system) characterize system behavior.

To find the ZIR for any proper rational system, follow the following steps:

1. find the denominator of $H(s)$
2. split 1/denominator into inverse-Laplace-transformable factors
3. inverse transform each part; this will only give you the *form* of the ZIR
4. actual ZIR is found by plugging in given initial conditions to get values for unknown constants

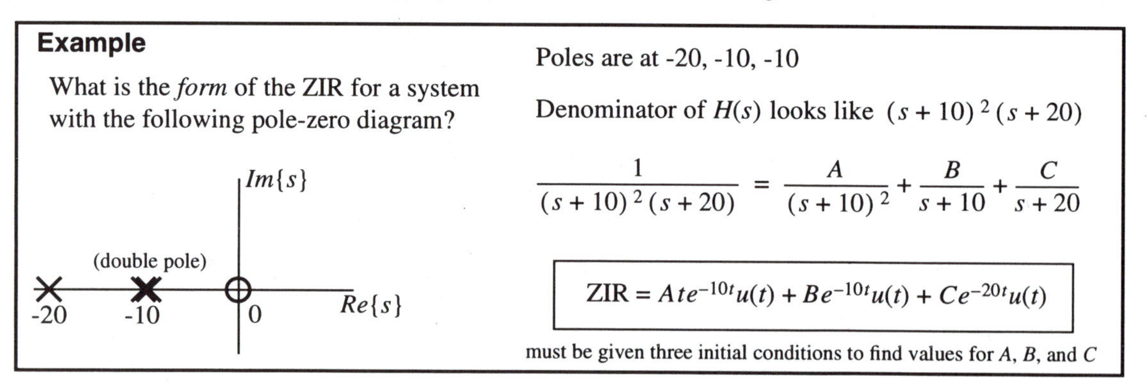

5.5 Zero State Response

The Zero State Response (ZSR) is the output of a system when presented with an input, assuming that all internal state variables are set to zero (i.e. the system is "at rest"). The most straightforward method of obtaining the ZSR (and thus solving the system's differential equation) is to use the forward and inverse Laplace transform in conjunction with the system function as follows:

Finding the ZSR

$$Y(s) = H(s)X(s)$$

$$y(t) = L^{-1}\{H(s)X(s)\} = \mathbf{ZSR}$$

Example

Given the system $\dot{y} + 3y = 5x$, find the output $y(t)$ when $x(t) = \sin(6t)\,u(t)$.

$$H(s) = \frac{5}{s+3} \qquad X(s) = \frac{6}{s^2+36}$$

$$Y(s) = \frac{30}{(s+3)(s^2+36)} = \frac{A}{s+3} + \frac{Bs+C}{s^2+36}$$

$$(A+B)s^2 + (3B+C)s + (36A+3C) = 30$$

$$A = \left.\frac{30}{s^2+36}\right|_{s=-3} = \frac{2}{3}$$

$$A + B = 0$$
$$3B + C = 0$$
$$36A + 3C = 30$$

$$A = \frac{2}{3},\ B = -\frac{2}{3},\ C = 2$$

$$Y(s) = \frac{2/3}{s+3} - \frac{(2/3)s}{s^2+36} + \frac{2}{s^2+36}$$

$$y(t) = \left(\frac{2}{3}e^{-3t} - \frac{2}{3}\cos 6t + \frac{1}{3}\sin 6t\right)u(t)$$

Note that it is possible to write the sum of several sinusoids of the same frequency as a single sinusoid with a new magnitude and phase. This can be accomplished by writing the sines and cosines in Euler form (see the Appendix) and recombining the complex exponentials. Therefore, an equivalent answer to this problem is:

$$y(t) = \left(\frac{2}{3}e^{-3t} + \frac{\sqrt{5}}{3}\cos(6t - 153.43°)\right)u(t)$$

Note that the e^{-3t} term in the ZSR looks like something that would also be part of the ZIR. This is not a coincidence! The input to the above system is a sine wave suddenly turned on at time=0. Such a sudden jolt to a system tends to set up some initial conditions that begin to decay away in the same manner that a ZIR would. After a long time, the input looks like a steady-state sine wave which then produces a steady-state output, just like an eigenfunction. The lesson to be learned here is that the ZSR will be the sum of a transient solution (which will have the same *form* as the ZIR) and a steady-state solution.

If there are any initial conditions present, the total solution can be found by adding together the ZIR and ZSR.

Total system response = ZIR + ZSR

5.6 The Impulse Response

The output of a system when the input is an impulse function $\delta(t)$ (defined in Chapter 9) has a special significance in linear system theory. We know that $Y(s)=H(s)X(s)$; if $x(t)=\delta(t)$ then $X(s)=1$ (take the Laplace transform) and thus we have $Y(s)=H(s)$. Recall from Section 5.1 that the function $H(s)$ completely characterizes a system. This must mean that the output $y(t)$ is now also a complete characterization of the system when the input is an impulse. We will designate this special output as $h(t)$, known as the impulse response of the system. $H(s)$, the system function, is the Laplace transform of $h(t)$. Slowly, everything is beginning to fit together. Chapter 10 contains a more detailed description of the impulse response and its uses.

It should also be apparent that the impulse response has the same *form* as the ZIR since it is simply the inverse transform of $H(s)$. A more intuitive explanation is that since the impulse exists only at t=0, its response can be thought of as instantaneously setting up some initial conditions and then letting them decay away at the natural frequencies of the system, just like a ZIR.

5.7 Combining Systems

The nice thing about systems is that they can be combined together to form bigger systems. The following diagrams illustrate the equivalent system function for two types of connections.

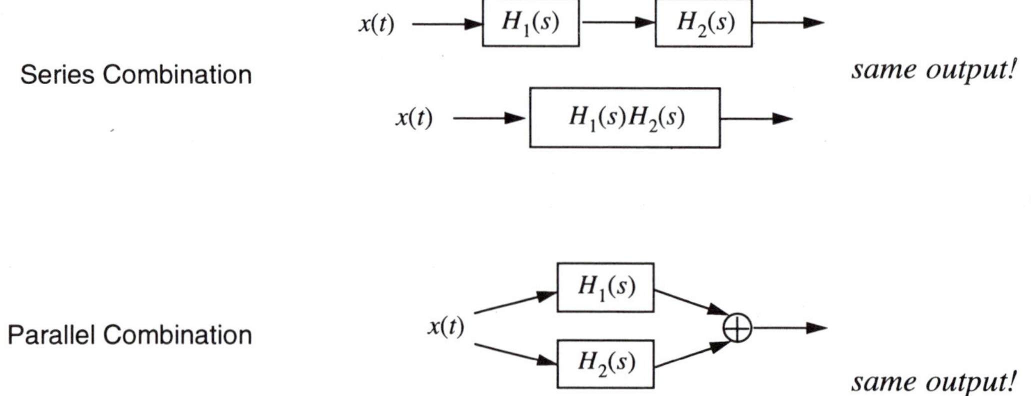

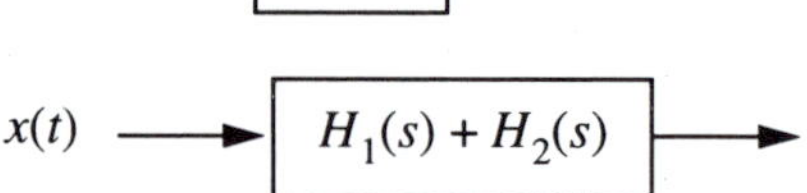

CHAPTER 6 Bode Plots

Overview

Bode plots are graphs of the steady-state response of stable, continuous-time LTI systems for sinusoidal inputs, plotted as change in magnitude and phase versus frequency. Bode plots are a visual description of a system; they are one of the most commonly used concepts from linear system theory. This chapter explains what a Bode plot is, steps through the mechanics of its computation, and then provides a set of rules for quickly graphing them. There is also a section on how to handle complex poles and zeros that discusses concepts like "Q" and resonance. Finally, there are examples of types of Bode plot problems and a discussion on how to determine a frequency response experimentally.

6.1 What is a Bode Plot?

A Bode plot, named after Dr. Bode, is the name given to a log-log plot of the *frequency response* of a continuous time system. So then what's a frequency response? It describes the steady-state response of a system to a sinusoid of a particular frequency. Recall that a sinusoidal function (complex exponential) is an eigenfunction of a CT system, meaning if $e^{j\omega t}$ goes in, $Ke^{j\omega t}$ comes out, where K is a complex number. A Bode plot is simply a graph of K vs. ω. Since K is complex, we will have to represent the frequency response as two sub-graphs. We could plot the real part and imaginary part separately, but it is much more intuitive to plot things as magnitude and phase versus ω. You can see this by rewriting K as $Ae^{j\phi}$. The output then becomes $Ae^{j(\omega t + \phi)}$, which has the same frequency as the input, but with a possible change in magnitude and phase of the sinusoid. In other words, $\sin(\omega t) \rightarrow A\sin(\omega t + \phi)$. Therefore, a Bode plot is simply a graph of A vs. ω and ϕ vs. ω. Simple, eh? Please note that things are plotted versus ω (radians/sec) not f (Hertz or cycles/sec). If needed, the conversion is $\omega = 2\pi f$.

The Big Picture

In general, e^{st} is an eigenfunction ($s = \sigma + j\omega$) of a CT systems as follows:

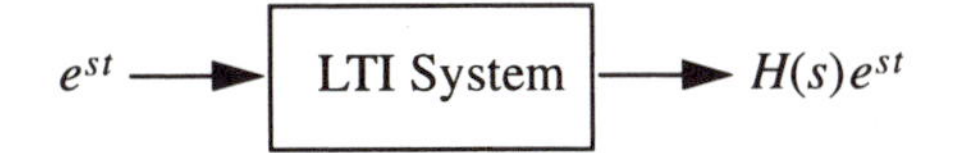

Here $H(s)$ is the familiar transfer function that we all know and love. So, from the previous discussion, you should realize that the Bode plot is nothing more than a graph of the transfer or system function $H(s)$ evaluated along the $j\omega$-axis (s=$j\omega$). In other words, the Bode plot is a representation of the transfer function of a continuous-time LTI system for steady-state sinusoidal inputs.

Bode plot = $H(s)$ evaluated along $j\omega$-axis (s=$j\omega$)

6.2 Calculating the Frequency Response

Let's illustrate the mechanics behind graphing a frequency response with a simple example. Given the $H(s)$ shown below, plug in different values of ω and compute the magnitude and phase of $H(j\omega)$.

$$H(s) = \frac{1}{s+10} \qquad H(j\omega) = \frac{1}{j\omega + 10}$$

$$\text{magnitude of } H(j\omega) = \frac{\text{magnitude of numerator}}{\text{magnitude of denominator}}$$

$$\text{phase of } H(j\omega) = \text{phase of numerator - phase of denominator}$$

$$\text{magnitude of } a + bj = \sqrt{a^2 + b^2}$$

ω	$\|H(j\omega)\|$	$\angle H(j\omega)$	$20\log_{10}\|H(j\omega)\|$
0	0.1000	0.0 deg	-20.00 dB
1	0.0995	-5.71 deg	-20.04 dB
2	0.0981	-11.31 deg	-20.17 dB
5	0.0894	-26.57 deg	-20.97 dB
10	0.0707	-45.00 deg	-23.01 dB
20	0.0447	-63.43 deg	-26.99 dB
50	0.0196	-78.69 deg	-34.15 dB
100	0.0100	-84.29 deg	-40.04 dB

You are probably used to plotting things on a linear scale, but something very interesting happens if you plot the magnitude of the frequency response on a logarithmic scale. Using a log scale for ω allows you to cover a wide range of possible input frequencies, and a log scale for the magnitude compresses big changes and emphasizes small changes to provide a better "big picture" of the overall response. The unit of choice for plotting log magnitude is the decibel, abbreviated dB. It is defined as $dB = 20\log_{10}amplitude$. Furthermore, the advantage of drawing things on a log-log scale is that the Bode plot becomes much easier to sketch by hand, as described in Section 6.3. The following two graphs show the above $|H(j\omega)|$ vs. ω on both linear and logarithmic axes.

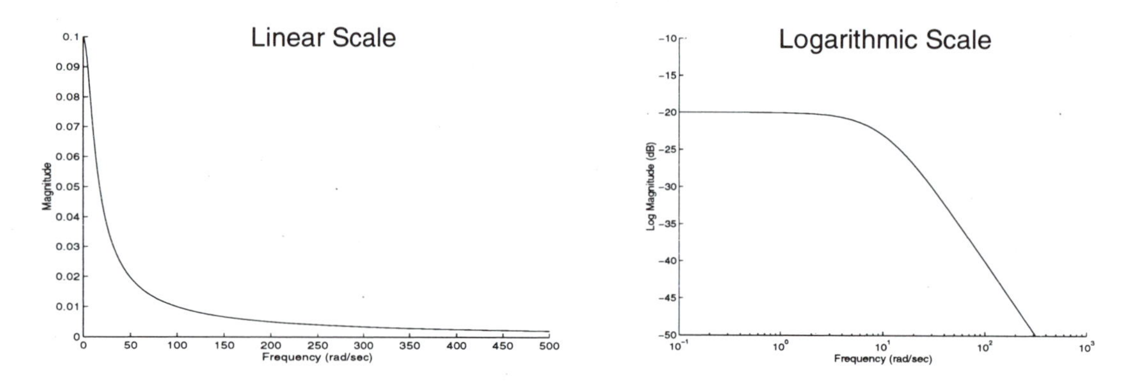

6.3 The Asymptotic Approximations

From examining the table and log-log plot in Section 6.2, it seems that there should be some straight line approximations to facilitate quick, but relatively accurate hand-drawn sketches of the Bode plot. Well, there are! The asymptotic magnitude plot only changes slope at a breakpoint (the location of a pole or zero). The phase plot changes slope at 0.1×breakpoint and at 10×breakpoint. At the breakpoint itself, the actual magnitude plot differs from the approximation by a factor of $20\log(1/\sqrt{2})$ or -3dB and the actual phase plot has changed by 45° and crosses the asymptotic approximation line. Note that these numbers scale for multiple order poles and zeros. For a single pole system, the breakpoint (3dB point) is also known as the half-power point since $|H(j\omega)|^2$ (power) is half as large. The asymptotic approximation technique is illustrated with the following example using the same $H(s)$ as above.

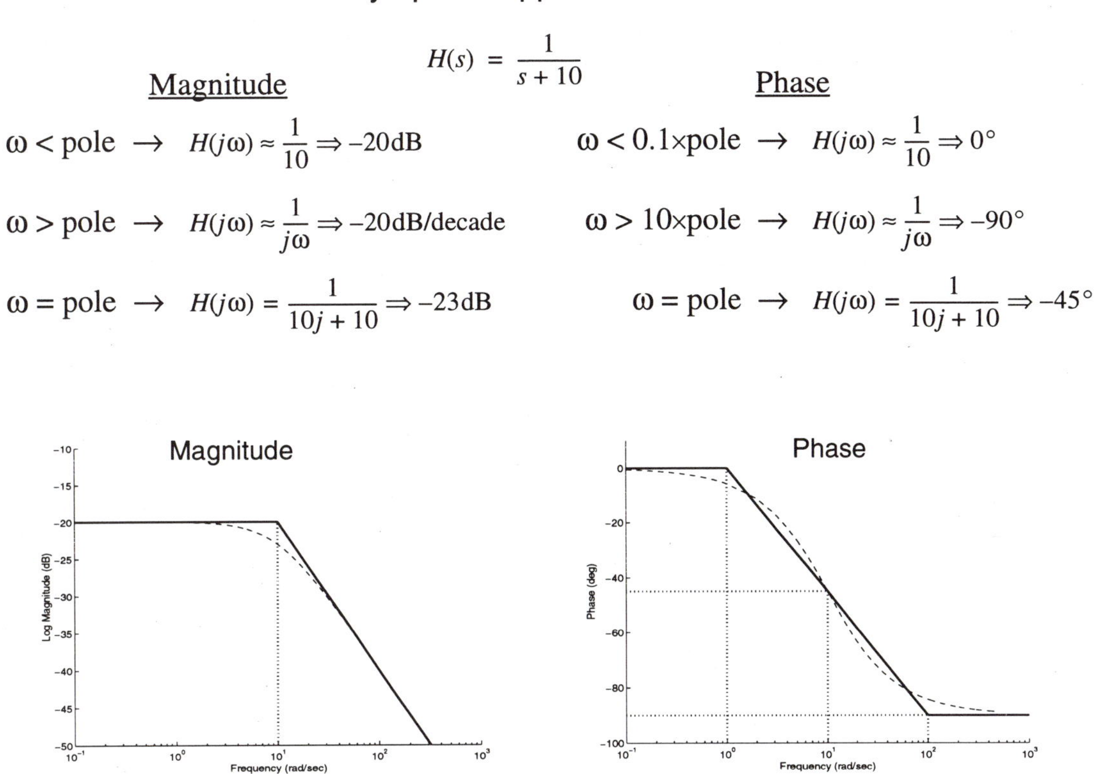

6.4 Relationship to the Pole/Zero plot

Since the Bode plot is essentially a graph of $H(s)$, we can easily relate it to the system pole/zero diagram. Imagine starting at $\omega=0$ and slowly sliding up the $j\omega$-axis. The changes in the length and angle of the vector from the pole/zero to the current location on the $j\omega$-axis are related to the changes in the magnitude and phase of the Bode plot. A common mistake is to say "Well, if the pole is at -10, then I should plug in -10 in for ω." No, no, no! You should plug in only positive values of ω when graphing the Bode plot. If the pole is at $s=-10$, then the breakpoint in the curve will occur at $\omega=10$ rad/sec. Between $\omega=0$ and $\omega=10$, the log-magnitude of the vector length stays roughly constant, which accounts for the flat portion of the Bode plot shown above. Hopefully the following diagram will make things clear.

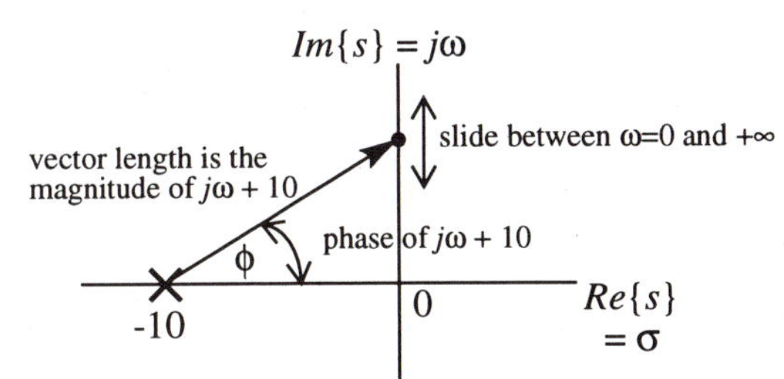

$$H(s) = \frac{1}{s+10} \qquad H(j\omega) = \frac{1}{j\omega+10}$$

As ω increases, $|j\omega + 10|$ increases $\Rightarrow$ $|H(j\omega)|$ decreases.

As ω increases, $\angle(j\omega + 10)$ goes to 90° $\Rightarrow$ $\angle H(j\omega)$ goes to -90°.

6.5 Dealing with Multiple Poles and Zeros

So far we have only sketched Bode plots with a single pole, but the methodology developed easily extends to systems with multiple poles and zeros. Since we are plotting log-magnitude, we can use the fact that the log of a product is the sum of the logs. And for phase, the phase of a product (division) is the sum (difference) of the phases. We illustrate this concept with the asymptotic approximations in the following example. Note that the sum of the asymptotic approximations starts to break down when the poles/zeros are too close together (within a factor of 10 or so).

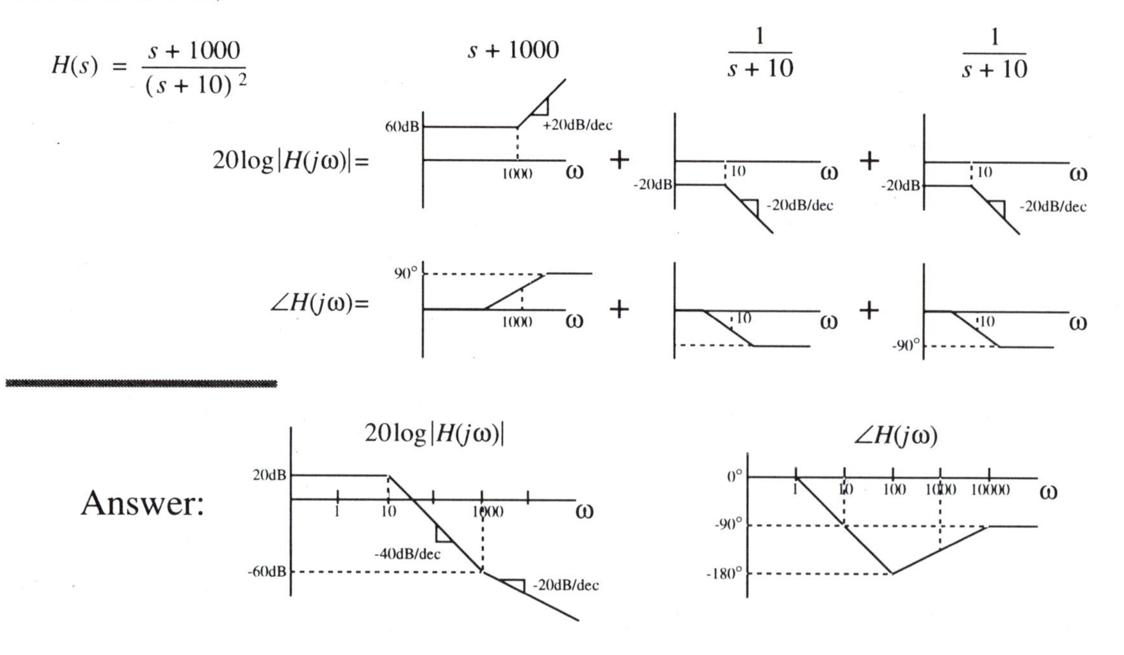

6.6 Summary of Bode Plotting Rules

The following set of rules applies to drawing Bode plots for systems with all poles and zeros on the real axis, spread sufficiently far apart (about a factor of 10 or more). Furthermore, we will assume that $H(s)$ is a causal, stable, minimum-phase system, meaning that all poles and zeros are in the left half plane. Causality and stability will be discussed in more detail in Chapter 14.

Plotting Magnitude $|H(j\omega)|$

1. Identify locations and order (how many) of all poles and zeros – these are the breakpoints.
2. Draw axes. Note that it is impossible to include ω=0 on a log scale. Start with a small ω, like 1, 0.00001, or whatever is appropriate. It is also useful to draw vertical dashed lines at breakpoints.
3. Starting at the left, the magnitude plot starts flat unless there is a pole at s=0 (start plot with slope of -20dB/dec for each pole at origin) or there is a zero at s=0 (start plot with slope of +20dB/dec for each zero at origin).
4. Continue drawing asymptote in a straight line until you reach a breakpoint (pole/zero).

5. For <u>each</u> pole encountered, *decrease* slope of asymptote by 20dB/dec. For <u>each</u> zero, *increase* slope by 20dB/dec. Go to step 4 unless there are no more breakpoints left.
6. Label one point on the y-axis by plugging in a value of ω into $H(j\omega)$ from any flat region of the plot. If there are no poles or zeros at the origin, using $\omega=0$ is the simplest choice. Continue labelling y-axis using slopes of asymptotes as guides.
7. Round corners inward by +/- 3dB for a more accurate magnitude plot.

Plotting Phase $\angle H(j\omega)$

1. Identify locations and order (how many) of all poles and zeros – these are the breakpoints.
2. Draw axes and vertical dashed lines at breakpoints.
3. Starting at the left, the phase plot starts at $\angle H(j\omega=0)$ (usually 0°). Plot starts at +90° for each zero at origin and -90° for each pole at the origin. A leading minus sign will add 180° to the phase. Plug in $j\omega$=very_small_imaginary_number and evaluate the phase manually if you're confused. Also remember, shifting the phase curve up or down by 360° doesn't change anything.
4. The phase plot continues as a flat line until reaching 0.1×breakpoint.
5. <u>Each</u> pole *subtracts* 90 degrees from the phase, spread over a distance of 0.1×pole location to 10×pole location. At the pole location, the phase has dropped by 45 degrees (halfway there). The situation is the same for zeros, but this time phase is *added*. Watch out for multiple poles/zeros. Go back to step 4 unless there are no more breakpoints left.
6. Round all corners to resemble an arctan curve (that's how phase is calculated) for more accurate plotting; the phase rounds by about 6° at 0.1×breakpoint and at 10×breakpoint.

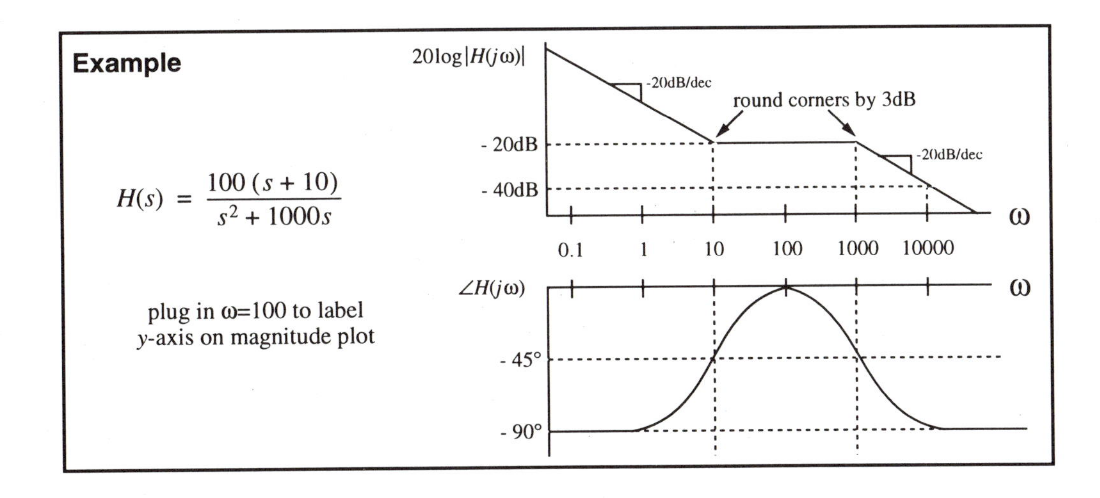

A Note about Slopes

A decade is defined to be a factor of 10 in frequency. An octave is defined as a factor of 2 in frequency (not a factor of 8). A sometimes useful conversion is 20 dB/decade ≈ 6 dB/octave. It can be derived as follows:

$$\text{dB difference in one octave} = 20\log(2x) - 20\log(x) = 20\log 2 + 20\log x - 20\log x = 20\log 2 = 6.02$$

6.7 Dealing with Complex Poles and Zeros

The name "Bode plot" is historically for plots of the frequency response of systems that have poles and zeros only on the real axis. But that doesn't mean you can't plot a frequency response for a system with complex poles/zeros. However, it is difficult to come up with a set of rules for drawing an accurate plot. If accuracy is needed, plug numbers into $H(s)$ or use MATLAB. Nevertheless, you should still be able to provide a very rough sketch of the frequency response, or at least be able to pick out the correct plot in a multiple choice question. Just to give you an idea of what these type of plots look like, here is an example of a Bode-like plot for a system with a pair of complex poles:

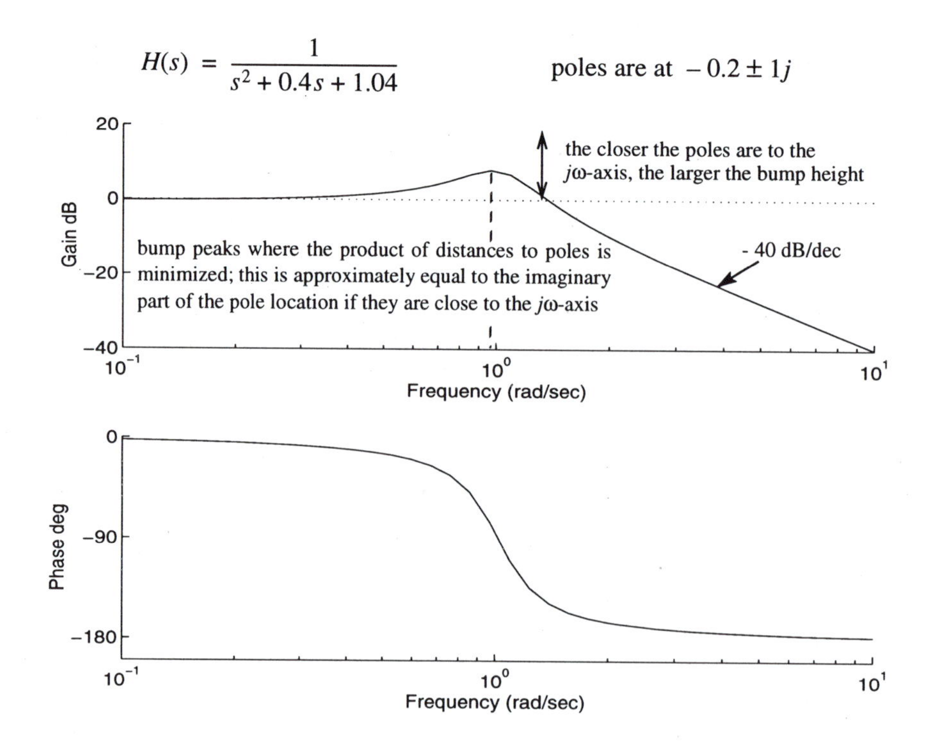

Sketching the Complex Pole/Zero Bode plot:

There are two ways of doing a "back-of-the-envelope" sketch of a complex pole/zero Bode plot. For both methods, first draw the pole/zero diagram. The first method is to observe the following exact relationship that is true when graphing any frequency response:

$$|H(j\omega)| = \prod_{i=1}^{numzeros} (\text{distance from zero}_i \text{ to } j\omega) \div \prod_{i=1}^{numpoles} (\text{distance from pole}_i \text{ to } j\omega)$$

$$\angle H(j\omega) = \sum_{i=1}^{numzeros} (\text{angle from zero}_i \text{ to } j\omega) - \sum_{i=1}^{numpoles} (\text{angle from pole}_i \text{ to } j\omega)$$

Trace your finger along the positive $j\omega$-axis and mentally approximate the above relations. "I'm getting closer to the pole, so the magnitude must be going up, etc." Note that all angles are measured relative to the positive real axis.

The second method is intimately related, but involves a more graphical approach. Look at the pole/zero diagram. Now, imagine placing a huge rubber sheet over the entire s-plane. Each pole represents a tent pole of infinite height, holding up the sheet. Each zero represents a nail, holding the rubber sheet to the ground. After you've supported/nailed down each pole/zero, you should be left with a circus tent-like structure. Now, remember that the Bode plot is $H(s)$ evaluated at $s=j\omega$? Well, the Bode magnitude plot is then the shape of your tent along the $j\omega$-axis! If there is a pole near the axis, the tent gets really high; likewise, if there is a zero near the axis, the tent is really low (if you take the logarithm, then it's really close to $-\infty$).

Another method of drawing the frequency response for complex poles and zeros is to draw what is known as a resonance curve. This is essentially the same as a Bode plot, but drawn using a linear scale on both axes. The resonant frequency is the input frequency that produces the maximum output. Sometimes resonance is good (like when trying to tune in to a particular radio station) or sometimes resonance is bad (like a car engine that starts vibrating only when it gets near a particular RPM). There are a few relationships that aid in a more accurate sketch of a resonance curve, especially when the poles are close to the $j\omega$-axis. For poles located at $-\alpha \pm \beta j$ with $\alpha \ll \beta$, the plot peaks at $\omega_{peak}=\beta$ and the half power bandwidth is 2α. Another way of describing a resonance curve is to define the sharpness of its peak, known as the "Q" of the system. A typical resonance curve is shown below:

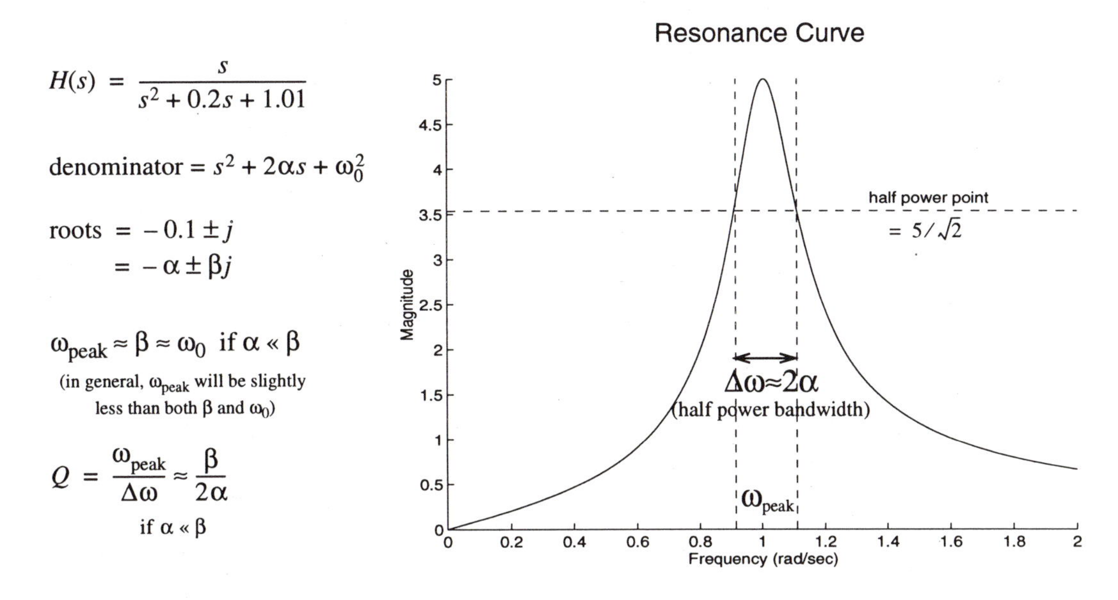

6.8 Sample Bode Plot Questions

There are generally four types of Bode plot problems:

1. Given an $H(s)$ with poles and zeros only on the real axis, accurately sketch its Bode plot. Note if you are just given the pole/zero diagram, you can still sketch the shape of a Bode plot, but the exact magnitude (y-axis values) will be impossible to determine (i.e. you can still say this point is 20dB lower than this other point, but the precise numerical values are unknown).
2. Given a Bode plot, recover the system function $H(s)$. This is a bit trickier; look for breakpoint locations and slopes of lines. Do not forget to look for a possible constant in front of $H(s)$; in other words, verify the y-axis values on the magnitude plot. Practice this with your friends! One person gets to practice drawing Bode plots, the other person practices reconstructing $H(s)$.

3. Multiple choice: given an $H(s)$, match it with the appropriate Bode plot (or vice versa). This type of question is especially common for systems with complex poles and zeros. Watch for subtleties like slopes of lines, y-axis values, locations of resonant peaks, etc.
4. You should not only know how to draw a Bode plot, but also be able to read one and understand what information it provides. An example problem is shown below:

What is the steady-state output of the following system? $H(s)$ and its Bode plot are given.

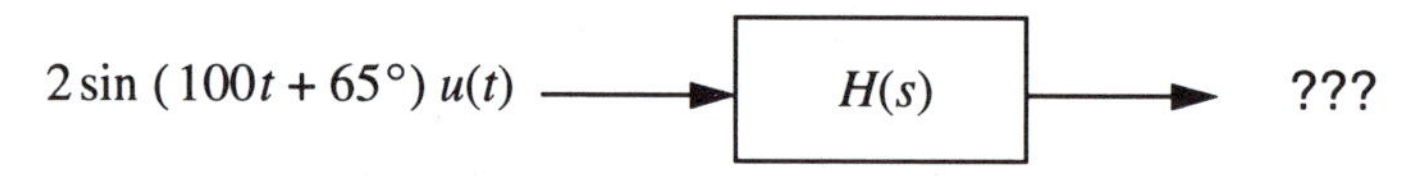

So, how would you do it?

Method 1: Take the Laplace transform of the input, multiply it by $H(s)$, take the inverse transform, and let $t \rightarrow \infty$. This method does work but is not suggested since it is unnecessarily too complicated. Note, if the input was $2\sin(100t + 65°)$ (without the $u(t)$), then the bilateral Laplace transform would not exist.

Method 2: Use steady-state analysis techniques, like shown in Section 3.6. Find the complex frequency of the input signal and plug it into $H(s)$, evaluating its magnitude and phase. Find the output signal by scaling the input by $|H(s)|$ and adding $\angle H(s)$ to the phase.

Method 3: Since we are dealing with a sinusoidal input and are looking for the steady-state output, we would do best by reading the answer directly off the Bode plot. Find the magnitude and phase response at ω=100 on the Bode plot. The output signal is then $2|H(100j)|\sin(100t + 65° + \angle H(100j))$.

6.9 Calculating the Frequency Response Experimentally

Suppose you've just designed the world's greatest stereo amplifier. You show it to your boss and she asks, "What's its frequency response?" Well, you could show her the Bode plot for the theoretical $H(s)$ using MATLAB, but that's not the real world. What can you do? Get yourself a function generator and an oscilloscope. Input a sinusoid of a particular frequency and display both the input and output sinusoids simultaneously on your scope. $|H(j\omega)|$ is the ratio of the output to input amplitudes and $\angle H(j\omega)$ is the phase difference between the output and the input. Note that the phase is negative when the output *lags* (is a delayed version of) the input. Continue recording these magnitudes and phase responses of your system for a suitable range of input frequencies. Remember, if you're trying to input a ω=1000 sine wave, that means you'll have to dial your function generator to 159.15 Hz ($\omega = 2\pi f$). A useful tip is that spacing input frequencies using the 1,2,5 method (1,2,5,10,20,50,100,200,500,1000, etc.) produces roughly equally spaced points on a logarithmic scale. Now plot your experimental results and show it to your boss!

CHAPTER 7 Discrete Signals and Z-Transforms

Overview

This chapter introduces the concept of a discrete-time signal. These signals consist of values that occur at discrete points in space or time, and are often samples of a continuous-time process. The Z-transform, a tool for discrete-time LTI system analysis is introduced. Several examples illustrating forward and inverse Z-transforms are provided, while paying particular attention to the notion of region of convergence.

7.1 A New Type of Signal

We now describe a new class of signals, known as discrete time (DT). These types of signals consist of values that occur at discrete points in space or time. The closing Dow Jones Industrial Average at the end of every day, the pixel values in an image of the moon, and the number of busses per hour going down Main Street are all examples of discrete signals. Discrete-time signals are not plotted versus time, but rather versus an index n, and are normally drawn as a series of "lollipops" with the height labeled.

How to denote a DT signal:

Some Basic DT Signals

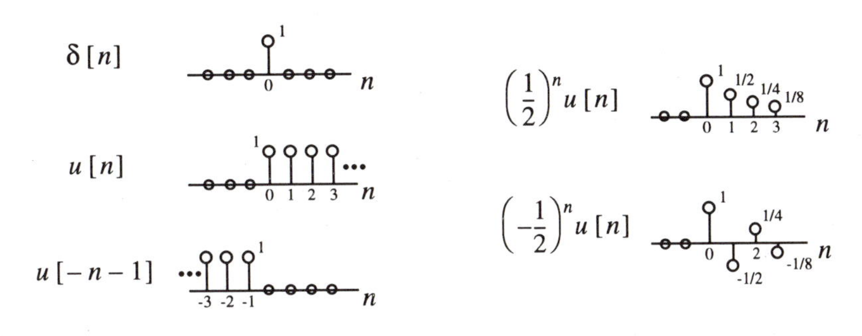

7.2 The Z-Transform

The discrete-time counterpart of the Laplace transform is the Z-transform. You can think of the Z-transform as simply an alternative way of expressing DT signals. Although initially it seems more complicated, writing discrete signals in the Z-domain greatly simplifies the relationship between inputs and outputs for signals passing through LTI systems.

$$\text{Definition:} \quad \tilde{X}(z) = \sum_{n=-\infty}^{\infty} x[n]\, z^{-n}$$

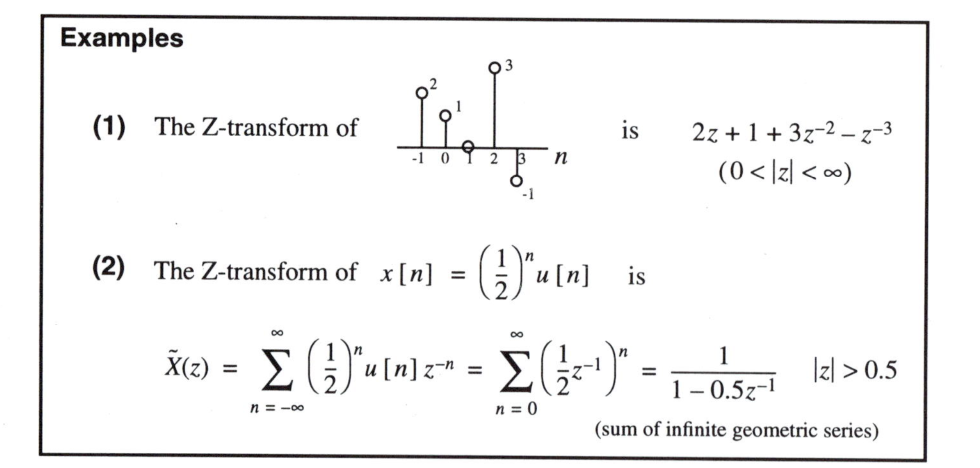

Examples

(1) The Z-transform of [stem plot: values 2, 1, 0, 3, -1 at n = -1, 0, 1, 2, 3] is $2z + 1 + 3z^{-2} - z^{-3}$ $(0 < |z| < \infty)$

(2) The Z-transform of $x[n] = \left(\frac{1}{2}\right)^n u[n]$ is

$$\tilde{X}(z) = \sum_{n=-\infty}^{\infty} \left(\frac{1}{2}\right)^n u[n]\, z^{-n} = \sum_{n=0}^{\infty} \left(\frac{1}{2} z^{-1}\right)^n = \frac{1}{1 - 0.5z^{-1}} \qquad |z| > 0.5$$

(sum of infinite geometric series)

7.3 Region of Convergence

Just like with Laplace Transforms, Z-transforms also have a region of convergence (ROC). The ROC defines the values of z for which the Z-transform sum will converge.

Note that the $j\omega$-axis in the s-plane maps to the unit circle in the z-plane using the transformation $z = e^{j\omega}$.

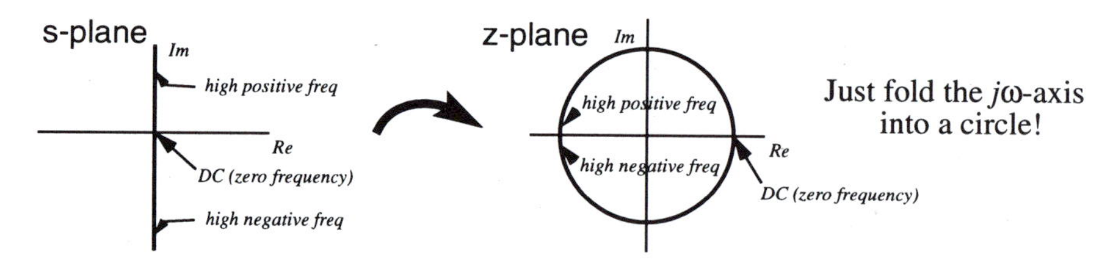

Since the ROC's for continuous-time signals look like half-planes or strips, the ROC's in the z-plane generally look like circles, holes, or donuts.

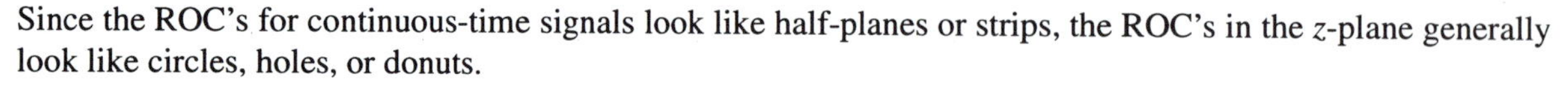

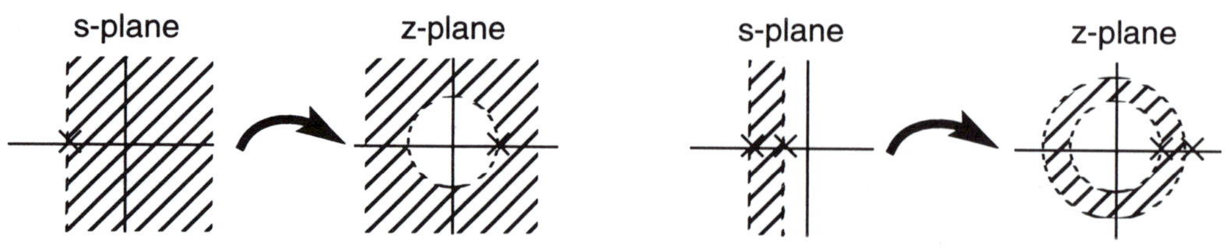

Types of Signals

The following definitions apply to both continuous and discrete time signals:

left-sided signal
(starts somewhere, ends at $-\infty$)

right-sided signal
(starts somewhere, ends at $+\infty$)
Note: a causal signal is one that is right-sided, but starts after 0

two-sided signal
(starts at $-\infty$, ends at $+\infty$)

finite-duration signal
(starts somewhere, ends somewhere)

Rules for the ROC

- The ROC never contains any poles (the values of z that make $\tilde{H}(z)=\infty$).
- If $x[n]$ is right-sided, the ROC is the area outside a circle, i.e. $|z| > a$, where a is the |outermost pole| .
- If $x[n]$ is left-sided, the ROC is the area inside of a circle, i.e. $|z| < a$, where a is the |innermost pole| .
- If $x[n]$ is two-sided or the sum of a left and right sided signal, the ROC is either a donut ($a < |z| < b$), or else the individual ROC's will not overlap, producing the null set.
- If $x[n]$ is of finite duration, then the ROC is the entire z-plane, except possibly z=0 and z=∞.

7.4 Z-Transform Pairs

Here are just a few sample Z-transform pairs: Note that the last two entries have different $x[n]$'s but appear to have the same $\tilde{X}(z)$'s. Looking closely however, the Z-transforms do differ in their region of convergence. Just as with Laplace transforms, the ROC will play a key role in performing the inverse transform (see Section 7.6).

$x[n]$		$\tilde{X}(z)$	*ROC*
$\delta[n]$	$\Leftrightarrow$	1	all z
$u[n]$	$\Leftrightarrow$	$\dfrac{1}{1-z^{-1}}$	$\|z\| > 1$
$a^n u[n]$	$\Leftrightarrow$	$\dfrac{1}{1-az^{-1}}$	$\|z\| > \|a\|$
$-a^n u[-n-1]$	$\Leftrightarrow$	$\dfrac{1}{1-az^{-1}}$	$\|z\| < \|a\|$

7.5 Z-Transform Properties

Property	*Time Domain*		*Z-Domain*	*New ROC*
Linearity	$ax[n] + by[n]$	$\Leftrightarrow$	$a\tilde{X}(z) + b\tilde{Y}(z)$	$\text{ROC} \supseteq \text{ROC}(x) \cap \text{ROC}(y)$
Time Shift	$x[n-k]$	$\Leftrightarrow$	$z^{-k}\tilde{X}(z)$	ROC(x) (but watch z=0, ∞)
Exponential Scaling	$a^n x[n]$	$\Leftrightarrow$	$\tilde{X}(z/a)$	$\{\lvert a\rvert \cdot z\}$ s.t. $z \in \text{ROC}(x)$
Linear Scaling	$nx[n]$	$\Leftrightarrow$	$-z\frac{d}{dz}\tilde{X}(z)$	ROC(x) (but watch z=0)
Time Reversal	$x[-n]$	$\Leftrightarrow$	$\tilde{X}(z^{-1})$	$\{1/z\}$ s.t. $z \in \text{ROC}(x)$

7.6 Inverse Z-Transform

The general formula for recovering $x[n]$ from $\tilde{X}(z)$ is the complex contour integral:

$$x[n] = \frac{1}{2\pi j}\oint \tilde{X}(z) z^{n-1} dz$$

DO NOT USE THIS FORMULA!!!

Rather, to perform the inverse Z-transform, we will merely manipulate the given expression until we see patterns we recognize from the Z-transform table. This heuristic scheme is just like the one used when doing inverse Laplace transforms.

Note, knowing the ROC is critical to performing an inverse Z-transform. For example:

$$\tilde{X}(z) = \frac{1}{1 - az^{-1}} \quad ? \quad \begin{cases} x[n] = a^n u[n] & \text{ROC: } \lvert z\rvert > \lvert a\rvert \\ x[n] = -a^n u[-n-1] & \text{ROC: } \lvert z\rvert < \lvert a\rvert \end{cases}$$

Is $x[n]$ $a^n u[n]$ or $-a^n u[-n-1]$? If we are told that the ROC is $\lvert z\rvert > \lvert a\rvert$, then we know that $x[n] = a^n u[n]$. Or equivalently, we could have been told that $x[n]$ was right-sided (or causal). Now for several examples:

Example

$\tilde{X}(z) = z^{-1} + 3 - 2z \qquad x[n] = ?$

By inspection, $x[n] = \delta[n-1] + 3\delta[n] - 2\delta[n+1]$

Example

$$\tilde{X}(z) = \frac{1-2z^{-1}+3z^{-2}}{1-z^{-1}}$$

Find $x[n]$, given that it is right-sided.

Method 1

$$\tilde{X}(z) = \frac{1}{1-z^{-1}} - \frac{2z^{-1}}{1-z^{-1}} + \frac{3z^{-2}}{1-z^{-1}} \quad \Rightarrow \quad x[n] = u[n] - 2u[n-1] + 3u[n-2]$$

Method 2

$$\begin{array}{r|l} & 1-z^{-1} \\ \hline 1-z^{-1} & 1-2z^{-1}+3z^{-2} \\ & \underline{1-z^{-1}} \\ & -z^{-1}+3z^{-2} \\ & \underline{-z^{-1}+z^{-2}} \\ & 2z^{-2} \end{array}$$

$$\tilde{X}(z) = 1 - z^{-1} + \frac{2z^{-2}}{1-z^{-1}}$$

$$\Rightarrow \quad x[n] = \delta[n] - \delta[n-1] + 2u[n-2]$$

1, -1, 2, 2, 2 ...
0 1 2 3 4 n

SAME ANSWER!

Example

$$\tilde{X}(z) = \frac{1}{1-\frac{5}{3}z^{-1}-\frac{2}{3}z^{-2}} \qquad \text{ROC: } \frac{1}{3} < |z| < 2$$

Find $x[n]$.

$$\tilde{X}(z) = \frac{1}{\left(1+\frac{1}{3}z^{-1}\right)(1-2z^{-1})} = \frac{A}{1+\frac{1}{3}z^{-1}} + \frac{B}{1-2z^{-1}}$$

If you have a hard time factoring $\tilde{X}(z)$, just look at the ROC. The boundaries of the ROC are the magnitude of the poles.

$$A = \tilde{X}(z)\left(1+\frac{1}{3}z^{-1}\right)\Big|_{z^{-1}=-3} \Rightarrow A = \frac{1}{7}$$

$$B = \tilde{X}(z)(1-2z^{-1})\Big|_{z^{-1}=\frac{1}{2}} \Rightarrow B = \frac{6}{7}$$

$$\tilde{X}(z) = \underbrace{\frac{1/7}{1+\frac{1}{3}z^{-1}}}_{\text{right-sided}} + \underbrace{\frac{6/7}{1-2z^{-1}}}_{\text{left-sided}}$$

How did we know which was left/right sided? Because the intersection of the two ROC's must be $1/3 < |z| < 2$

$$\Rightarrow \quad x[n] = \frac{1}{7}\left(-\frac{1}{3}\right)^n u[n] - \frac{6}{7}(2)^n u[-n-1]$$

7.7 Initial and Final Value Theorems

Just as with continuous time signals, one can also determine the initial and final value of a discrete time signal by examining only its Z-transform. Note that the use of the initial value theorem assumes that $x[n]=0$ for $n<0$.

Initial Value Theorem

$$x[0] = \lim_{z \to \infty} \tilde{X}(z)$$

Final Value Theorem

$$\lim_{n \to \infty} x[n] = \lim_{z^{-1} \to 1} (1 - z^{-1})\, \tilde{X}(z)$$

CHAPTER 8 Discrete-Time Systems

Overview

The analysis techniques for discrete-time systems are very similar to those used for CT systems. Some ways of describing a discrete-time system include difference equations, system functions $\tilde{H}(z)$, and delay-adder-gain block diagrams. The concept of a frequency response for a DT system is also discussed. Finally, methods for converting between continuous-time systems and discrete-time systems are provided.

8.1 Difference Equations and the System Function

Continuous-time systems are described by *differential* equations. Discrete-time LTI systems are described by linear constant-coefficient *difference* equations such as:

$$y[n] + \frac{1}{6}y[n-1] - \frac{1}{6}y[n-2] = x[n] + 2x[n-1]$$

It is possible to convert this difference equation to a system function representation by taking the Z-transform of both sides and utilizing the time-delay property.

$$\tilde{Y}(z) + \frac{1}{6}z^{-1}\tilde{Y}(z) - \frac{1}{6}z^{-2}\tilde{Y}(z) = \tilde{X}(z) + 2z^{-1}\tilde{X}(z)$$

$$\tilde{H}(z) = \frac{\tilde{Y}(z)}{\tilde{X}(z)} = \frac{1 + 2z^{-1}}{1 + \frac{1}{6}z^{-1} - \frac{1}{6}z^{-2}} \qquad \text{ROC is } |z| > \frac{1}{2}$$

(assuming causal)

$$\boxed{\tilde{Y}(z) = \tilde{H}(z)\tilde{X}(z)}$$

$\tilde{H}(z)$, together with its ROC, is a complete characterization of this discrete-time LTI system. Just like in continuous-time, the poles and zeros of the system are those values of z that make the denominator and numerator go to zero. $h[n]$ is the inverse Z-transform of $\tilde{H}(z)$ and is known as the impulse response of a DT system.

Complex exponentials of the form z^n are eigenfunctions of discrete-time LTI systems. That is, if z^n goes in, $\tilde{H}(z)z^n$ comes out. For example:

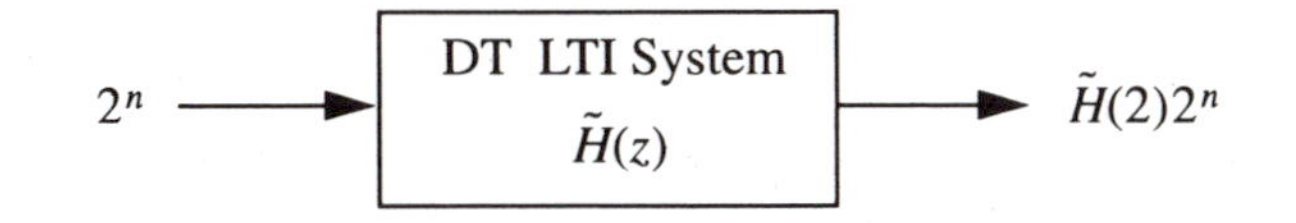

8.2 Discrete-Time Frequency Response

In Chapter 6 we plotted the Bode plot or frequency response of a continuous-time system by evaluating the magnitude and phase of $H(s)$ along the $j\omega$-axis. Recall from Section 7.3 that the $j\omega$-axis maps to the unit circle in the z-plane. Therefore, it is possible to plot the frequency response of a DT system by evaluating the magnitude and phase of $\tilde{H}(z)$ along the unit circle, which means $z = e^{j\omega}$ for $0 \leq \omega < 2\pi$. Just like with Bode plots, the following rules hold:

$$|\tilde{H}(e^{j\omega})| = \prod_{i=1}^{numzeros} (\text{distance from zero}_i \text{ to } e^{j\omega}) \div \prod_{i=1}^{numpoles} (\text{distance from pole}_i \text{ to } e^{j\omega})$$

$$\angle\tilde{H}(e^{j\omega}) = \sum_{i=1}^{numzeros} (\text{angle from zero}_i \text{ to } e^{j\omega}) - \sum_{i=1}^{numpoles} (\text{angle from pole}_i \text{ to } e^{j\omega})$$

Note that all angles are measured relative to a horizontal vector pointing to the right. For a far more detailed description of discrete-time frequency response as well as several examples, see Section 5.3 of *Discrete-Time Signal Processing* by Oppenheim and Schafer (1989).

8.3 Delay-Adder-Gain Block Diagrams

Discrete-time systems are sometimes described graphically by delay-adder-gain block diagrams. After manipulating the difference equation into a standard form, the block diagram can be easily drawn by simply filling in the appropriate coefficients in the template shown below. This canonical form of the delay-adder-gain block diagram is derived on pages 216-218 of *Circuits, Signals, and Systems* by Siebert (1986).

$$\sum_{k=0}^{N} a_k y[n-k] = \sum_{k=0}^{N'} b_k x[n-k]$$

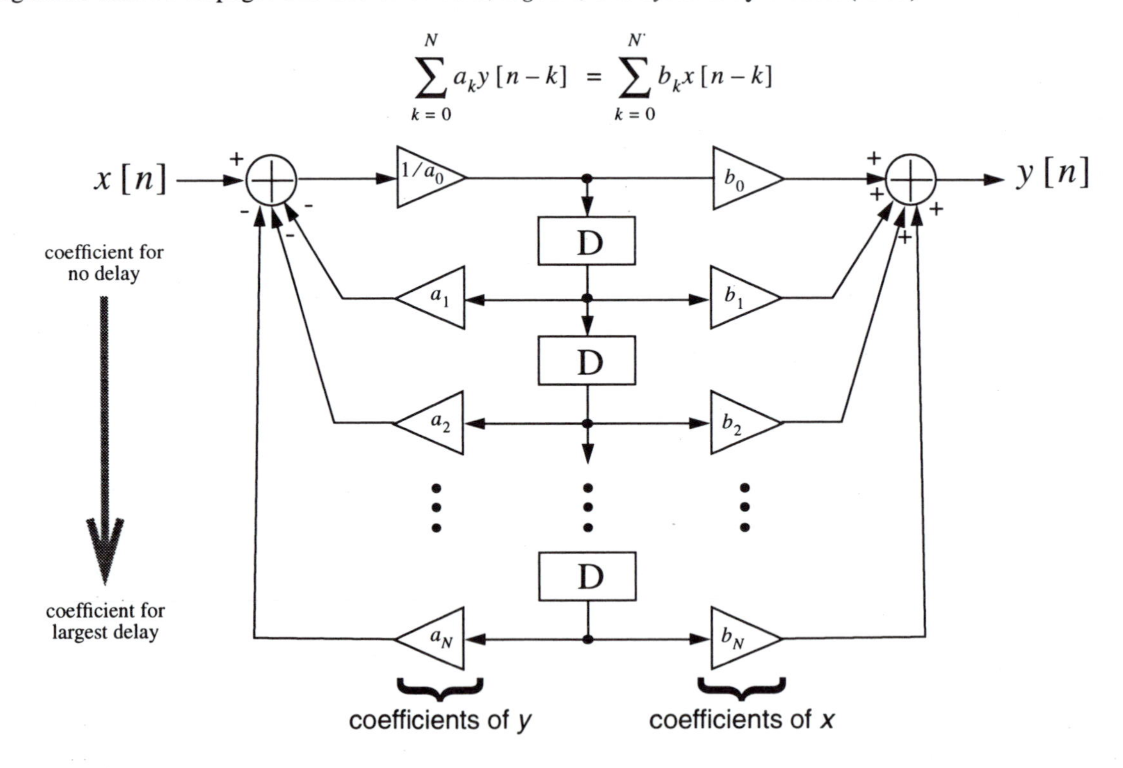

8.4 Describing a System

Each of the following is an equivalent way to describe a continuous-time or discrete-time system. Given any one of them, it is possible to reproduce the other four. Some of these concepts have not been fully introduced yet, but are shown here just so that you can begin thinking about how they are interrelated.

- the differential or difference equation (and whether the output is left, right, or two-sided)
- the system function $H(s)$ or $\tilde{H}(z)$ and its ROC
- the impulse response $h(t)$ or $h[n]$
- the step response
- the integrator or delay-adder-gain block diagram

Also, given just the pole-zero diagram and the ROC you should be able to recreate any of the previous five items, but only within a constant factor.

8.5 Converting CT Systems to DT Systems

There are a variety of ways to convert a continuous-time system to a discrete-time system. The impulse invariance method is based on sampling the CT impulse response. Other methods are based upon discrete approximations of the derivative. Some of these transformations are summarized below and fully described in Section 10.8 of *Signals and Systems* by Oppenheim et al (1983). Just replace s or z in the system function with the appropriate expression to convert from CT to DT or vice versa. The variable T is the length of time between sample points. Please note that these are all *approximations*, each having its own limitations.

Impulse Invariance	$z = e^{sT}$	
Backward-Difference	$s = \dfrac{1-z^{-1}}{T}$	$z = \dfrac{1}{1-sT}$
Bilinear Transformation	$s = \left(\dfrac{2}{T}\right)\dfrac{1-z^{-1}}{1+z^{-1}}$	$z = \dfrac{1+(T/2)\,s}{1-(T/2)\,s}$

8.6 Combining DT Systems

Discrete-time systems combine together in the same manner as continuous-time systems. The following diagram illustrates the equivalent system functions for serial and parallel connections.

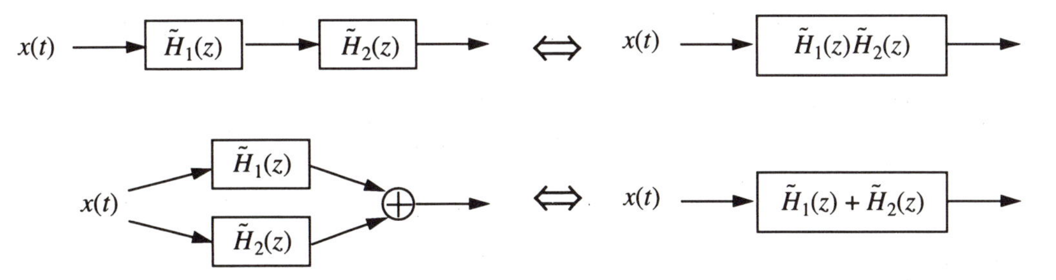

CHAPTER 9 Generalized Functions

Overview

This chapter defines and describes the behavior of some basic, but critical continuous-time and discrete-time functions used in LTI system analysis.

9.1 The Impulse

The unit impulse, denoted by $\delta(t)$, is nothing more than a mathematical entity that, like so many other academic things, does not exist in the real world. However, it will prove immensely useful for the analysis of continuous and discrete-time systems. The graphical definitions for both continuous-time and discrete-time unit impulses are shown below. Note that in the CT case, the impulse has infinite *amplitude* but finite *area*, since it has infinitesimal width. It is convention to write the area of the impulse next to the arrowhead, with the length of the arrow proportional to the impulse area. As far as the real world is concerned, you can consider the impulse to be a very, very short pulse; for example, a pulse of laser light or the force when hitting a baseball with a bat can be modeled with impulses. Since the relationship between an impulse and a short pulse is known in both the time and frequency domains (see Chapter 16), it is easy to later correct/modify your original analysis to account for these real world issues.

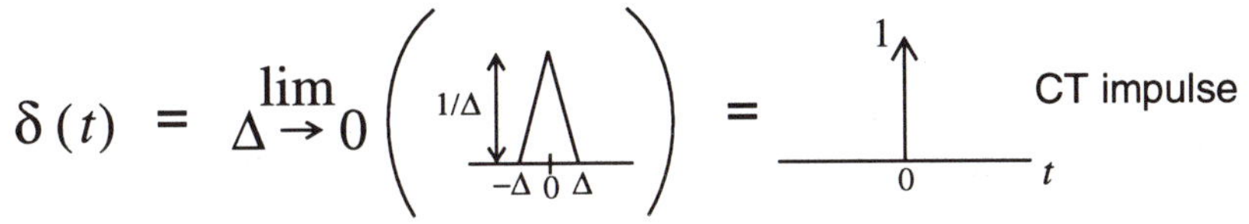

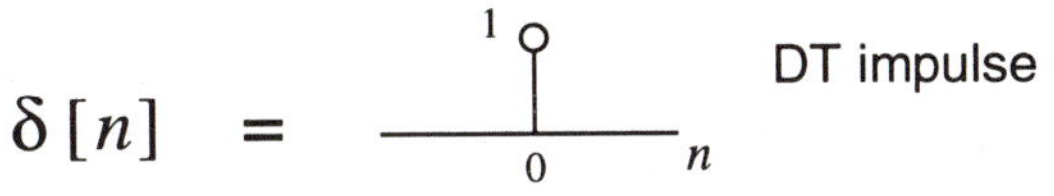

9.2 Derivatives of Discontinuities

The impulse can be used to represent the derivative of a function at a discontinuity. The area of the impulse is equal to the height of the discontinuity. Be sure to watch for the correct sign. An example:

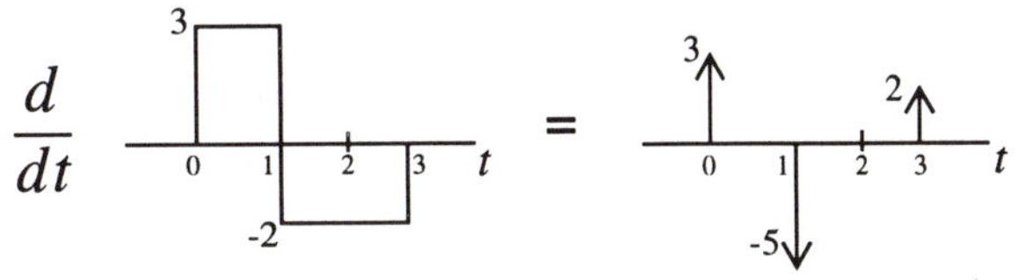

9.3 The Doublet

The doublet is an extension of the definition of the impulse function. It has uses as a differentiation operator, as will become clear later. Note that the continuous time doublet is the derivative of the unit impulse.

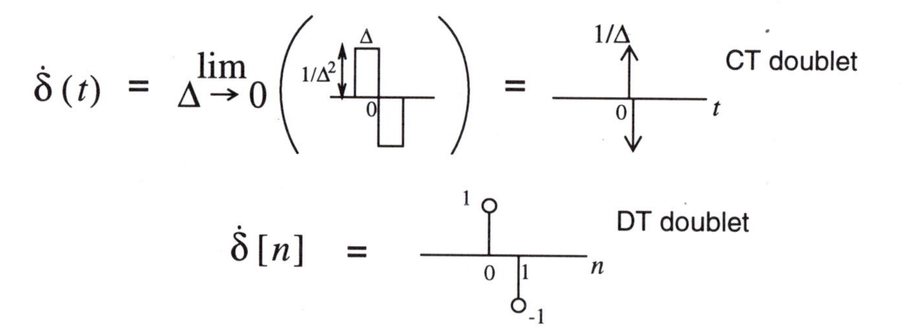

9.4 Step Functions

The continuous and discrete-time unit step functions $u(t)$ and $u[n]$ are defined as shown below:

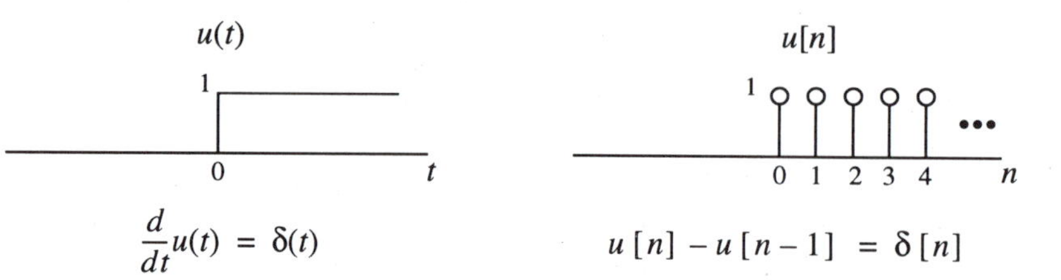

$$\frac{d}{dt}u(t) = \delta(t) \qquad u[n] - u[n-1] = \delta[n]$$

9.5 Properties

Generalized functions like the impulse and doublet are best described not by what they "are", but what they "do":

$$x(t)\delta(t-t_0) = x(t_0)\delta(t-t_0)$$

Multiplying a time function by a unit impulse produces an impulse with area given by the height of the time function at the location of the impulse.

$$x(t_0) = \int_{-\infty}^{\infty} x(t)\delta(t-t_0)dt$$

Multiplying by a unit impulse and *integrating* picks out the value of the original function at the location of the impulse.

$$-\dot{x}(t_0) = \int_{-\infty}^{\infty} x(t)\dot{\delta}(t-t_0)dt$$

Multiplying by a doublet and integrating picks out the *negative* of the derivative of the original function at the location of the doublet because the doublet acts like taking $(x(t_0) - x(t_0+\Delta))/\Delta$.

$$\delta(at) = \frac{1}{|a|}\delta(t)$$

Time scaling a CT impulse is the same as changing its area.

$$u(t) = \int_{-\infty}^{t} \delta(\tau)d\tau$$

The unit step function is the integral of the delta function.

$$\delta(t) = \frac{d}{dt}u(t)$$

The delta function is the derivative of the unit step function.

CHAPTER 10 The Impulse Response and Convolution

Overview

The impulse response $h(t)$ is one of the fundamental concepts of LTI system theory. It is a complete characterization of a system and is intimately related to $H(s)$, the system function. This chapter defines the impulse response and derives the convolution operation, a tool for the time-domain analysis of systems. Finally, the link between time domain and frequency domain analysis is complete. The chapter concludes with several properties of the convolution operator, which provide further insight and tools for block diagram analysis and simplification.

10.1 CT and DT Signals are Made of Impulses

All continuous-time and discrete-time signals can be represented as the sum of scaled and shifted unit impulses. This is fairly obvious for discrete signals by rewriting $x[n]$ as:

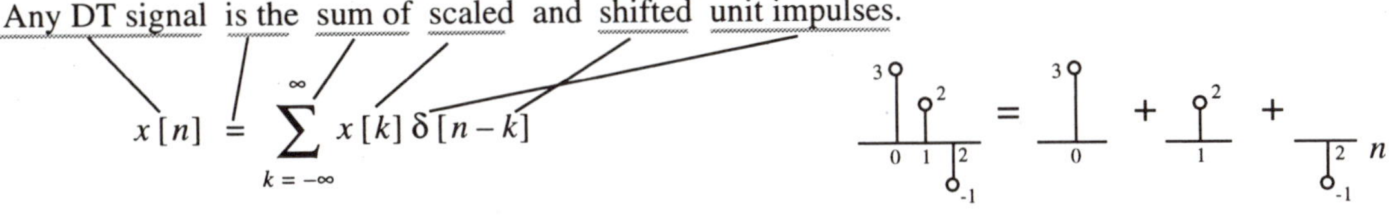

$$x[n] = \sum_{k=-\infty}^{\infty} x[k]\,\delta[n-k]$$

In continuous-time, it's a little harder to see, but the concept is the same: $x(t) = \int_{-\infty}^{\infty} x(\tau)\delta(t-\tau)d\tau$

See Section 3.1 of *Signals and Systems* by Oppenheim et al (1983) for its derivation.

10.2 Definition of Impulse Response

We will now define $h(t)$ and $h[n]$ to be the output of an LTI system when the input is a unit impulse, hence the name *impulse response*. Since the system is linear and time-invariant, the response to any size impulse located anywhere in time is obtained by merely scaling and shifting $h(t)$ or $h[n]$. Now, since we have shown that a signal can be broken down into a set of impulses, the response of an LTI system to an arbitrary input signal is simply the sum of its scaled and shifted impulse responses. This concept leads to the time-domain technique known as *convolution*. Note that it is improper to discuss an impulse response for non-LTI systems.

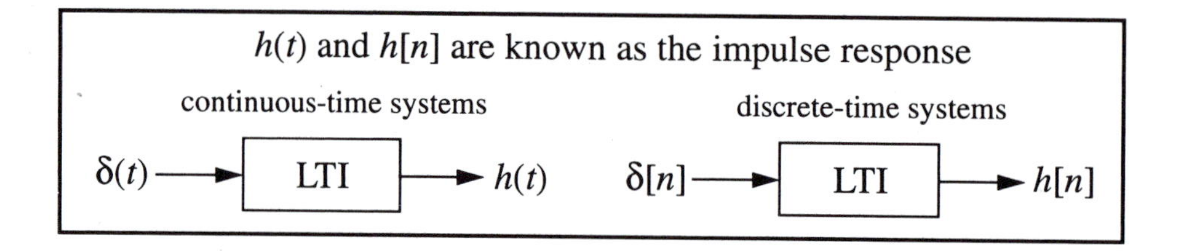

10.3 The Convolution Integral and Sum

The impulse response $h(t)$ or $h[n]$ is sufficient to completely characterize an LTI system. Once it is known, the system's response to any input can be found through the convolution formulas given below.

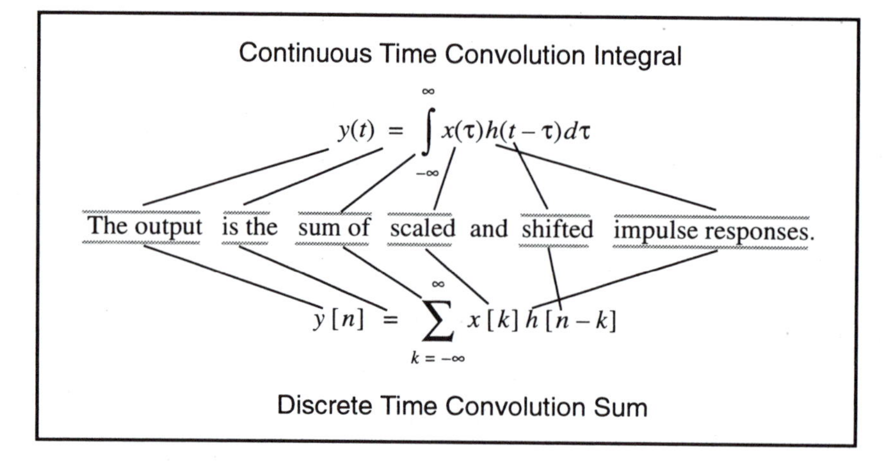

It is common to name systems (i.e. blocks) in block diagrams by their impulse response, as shown below. The asterisk ($*$) is used to symbolize the convolution operation.

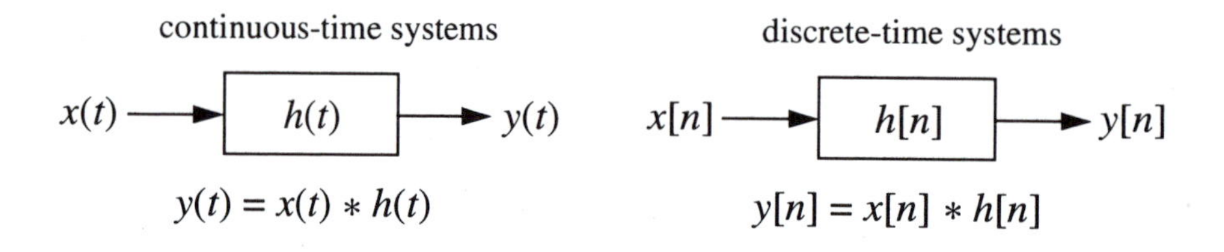

$y(t) = x(t) * h(t)$ $\qquad$ $y[n] = x[n] * h[n]$

10.4 Eigenfunctions Revisited

We first introduced the concept of eigenfunctions in Section 3.3. Now we have the tools necessary to demonstrate why complex exponentials of the form e^{st} are indeed eigenfunctions of continuous-time systems.

continuous-time convolution formula	$y(t) = \int_{-\infty}^{\infty} h(\tau)x(t-\tau)d\tau$
substitute in $x(t) = e^{st}$ input	$y(t) = \int_{-\infty}^{\infty} h(\tau)e^{s(t-\tau)}d\tau$
move e^{st} outside the integral	$y(t) = e^{st}\int_{-\infty}^{\infty} h(\tau)e^{-s\tau}d\tau$
use definition of Laplace transform	$y(t) = H(s)e^{st}$

10.5 The Link Between Time Domain and Frequency Domain

The convolution operation is our tool for the time-domain analysis of systems. Like everything else in the world of signals and systems, it fits quite nicely into the big picture. Our old friend $H(s)$ the system function is simply the Laplace transform of the impulse response $h(t)$. Similarly, $\tilde{H}(z)$ is the Z-transform of $h[n]$. Take a moment to allow that to sink in.

This momentous conclusion allows us to derive the relationship between the time-domain and frequency-domain analysis of systems. Consider an LTI system with input $x(t)$, impulse response $h(t)$, and output $y(t)$.

convolution operation	$y(t) = \int_{-\infty}^{\infty} x(\tau)h(t-\tau)d\tau$
take Laplace transform	$Y(s) = L\{y(t)\} = \int_{-\infty}^{\infty}\int_{-\infty}^{\infty} [x(\tau)h(t-\tau)d\tau]\, e^{-st}dt$
exchange order of integration	$Y(s) = \int_{-\infty}^{\infty} x(\tau)\left[\int_{-\infty}^{\infty} h(t-\tau)e^{-st}dt\right]d\tau$
use time-shift property	$Y(s) = \int_{-\infty}^{\infty} x(\tau)H(s)e^{-s\tau}d\tau = H(s)\int_{-\infty}^{\infty} x(\tau)e^{-s\tau}d\tau$
def. of Laplace transform	$Y(s) = H(s)X(s)$

So from the above derivation, we can arrive at the following conclusion:

$$y(t) = h(t) * x(t) \Leftrightarrow Y(s) = H(s)X(s) \qquad \text{convolution in time = multiplication in frequency}$$

Now, finally things are beginning to come full circle. Convolution provides the link between the time-domain and frequency-domain analysis of LTI systems. By now, you should realize that there are two ways to answer the question: "What is the output of the system whose input is $x(t)$ and whose impulse response is $h(t)$?"

1. Convolve† $x(t)$ and $h(t)$ directly to find $y(t)$ using mathematical or graphical means. The mechanics of continuous time and discrete time convolution are described in the following two chapters.
   †The verb form of the word is "convolve", not "convolute" as many students have mistakenly said.
2. Find $X(s)$ and $H(s)$ by taking Laplace (or Fourier - see Chapter 16) transforms, multiply them together, and then take the inverse transform to get back to $y(t)$. Although seemingly longer, the frequency domain method is often easier to compute. However, always keep in mind the direct method of convolution in the time domain; it provides insight often lost with the frequency domain method and will come in quite handy sometimes.

10.6 Relation to Step Response

Some other fields – especially control theory – emphasize the *step response* of a system, which we will denote as $s(t)$ or $s[n]$. The step response is, as expected, the output of the system when the input is a unit step. Note that the unit step is the integral of the impulse, and convolution is a linear operation. This implies that the step response is the integral of the impulse response. In general, the relationship between the two is as follows:

$$s(t) = u(t) * h(t) \qquad s[n] = u[n] * h[n]$$

$$s(t) = \int_{-\infty}^{t} h(\tau)d\tau \qquad s[n] = \sum_{k=-\infty}^{n} h[k]$$

$$h(t) = \frac{d}{dt}s(t) \qquad h[n] = s[n] - s[n-1]$$

10.7 Properties of Convolution

Some mathematical properties of the convolution operation which are critical to LTI systems analysis and useful for simplifying block diagrams are shown below:

Linearity

Convolution is a linear operator. Any linear operation on either $x(t)$ or $h(t)$ will produce the same linear operation on the output. In other words, doubling the input will double the output; taking the derivative of $h(t)$ will produce the derivative of the output, etc.

Time-Invariance

Any shifts in time of either $x(t)$ or $h(t)$ will produce the corresponding shift in the output. In other words, delaying the input will delay the output the same amount.

Commutative

The order in which convolution is performed doesn't matter: $x(t) * h(t)$ is the same as $h(t) * x(t)$. This means you could pass $x(t)$ through a system with impulse response $h(t)$ or pass $h(t)$ through a system with impulse response $x(t)$ and still get the exact same answer. Another interpretation is:

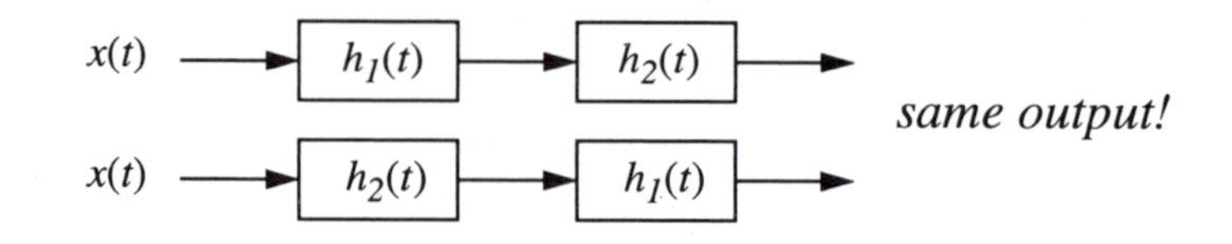

Associative

The manner in which convolution operations are grouped doesn't matter. Systems (blocks) in series can be combined together through convolution. $x(t) * [h_1(t) * h_2(t)] = [x(t) * h_1(t)] * h_2(t)$. A graphical interpretation:

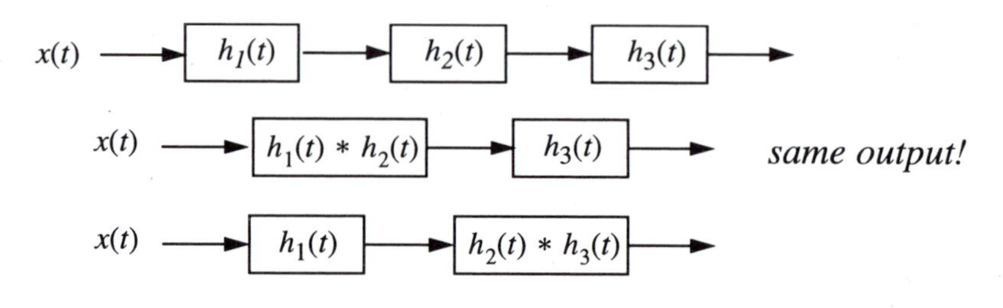

Distributive

Systems along parallel paths can be combined through addition. $x(t) * [h_1(t) + h_2(t)] = [x(t) * h_1(t)] + [x(t) * h_2(t)]$. A graphical interpretation:

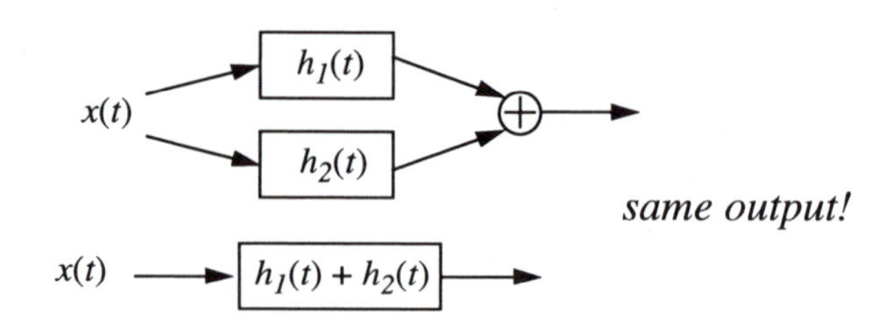

CHAPTER 11 Discrete-Time Convolution

Overview

This chapter describes the mechanics of discrete-time convolution. A step-by-step procedure derived directly from the convolution sum is explained and illustrated with examples. Other means of convolution explored include the signal decomposition method and directly summing the equation in the case of infinite length signals.

11.1 Graphical Flip/Shift Method

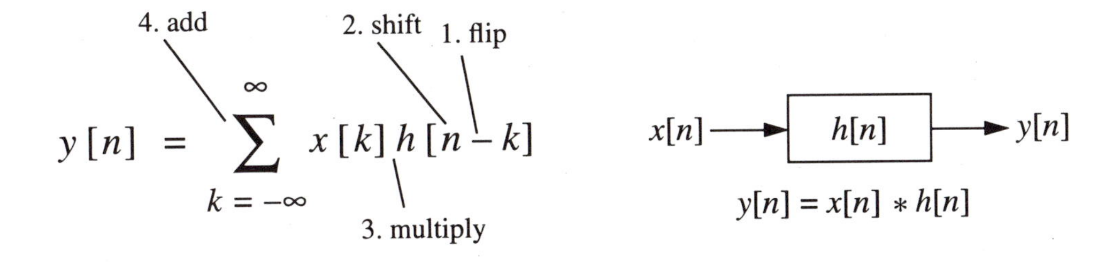

Looking at the equation for convolution, we can arrive at the following step-by-step procedure:

1. Choose one signal to be $x[n]$, the other is then $h[n]$; draw them both on the k axis.
2. *FLIP* $h[k]$ about k=0.
3. *SHIFT* flipped version of h to the right by n.
4. *MULTIPLY* $x[k]$ by the flipped/shifted version of $h[k]$ and *ADD* across all values of k.
5. The summation in step 4 gives you $y[n]$ for only one value of n.
6. Repeat steps 3-5 for all possible values of n.

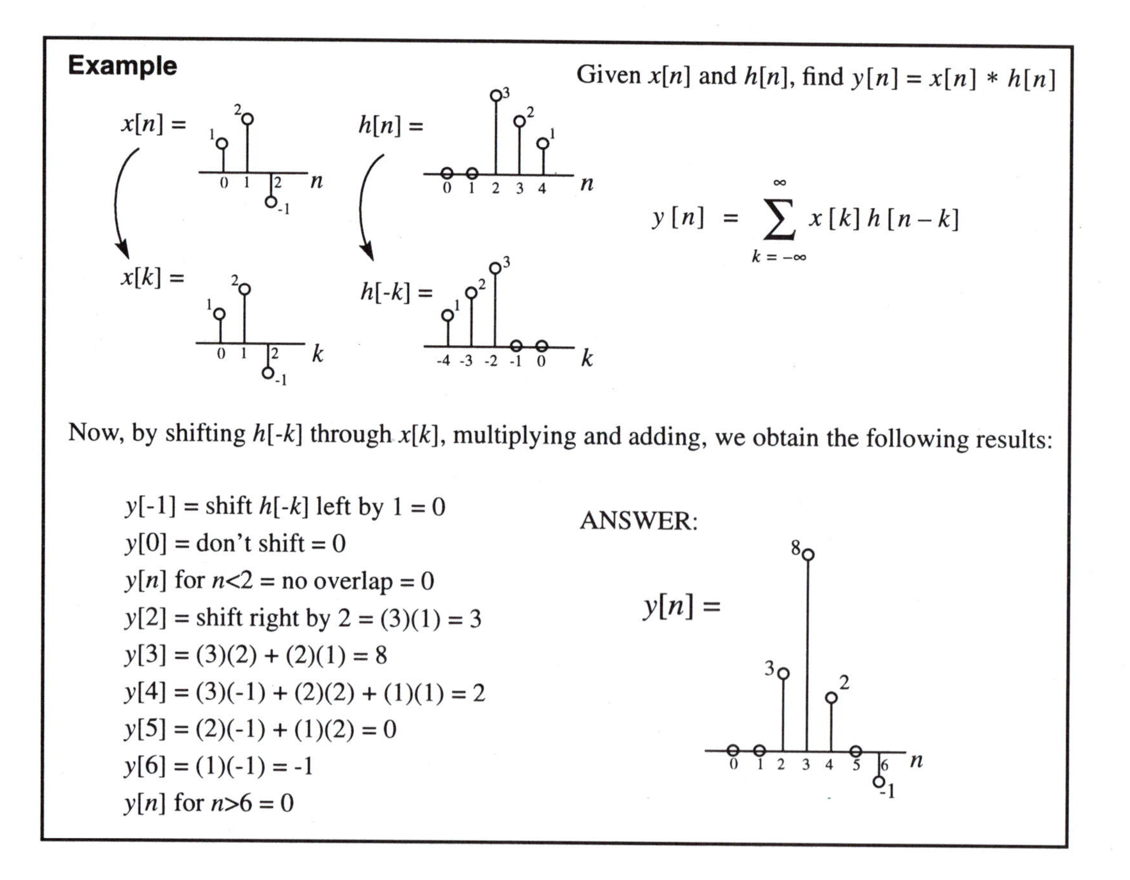

11.2 Convolving with Impulses

Convolving any signal with an impulse $\delta[n]$ produces the same signal at the output, i.e. $x[n] * \delta[n] = x[n]$. Furthermore, if the impulse is not at n=0 or of unit height, the result of $x[n] * A\delta[n-k]$ is simply $Ax[n-k]$. In other words, the output is simply a shifted and scaled version of $x[n]$. If you have any doubts, I suggest verifying the following example using the complete flip/shift method.

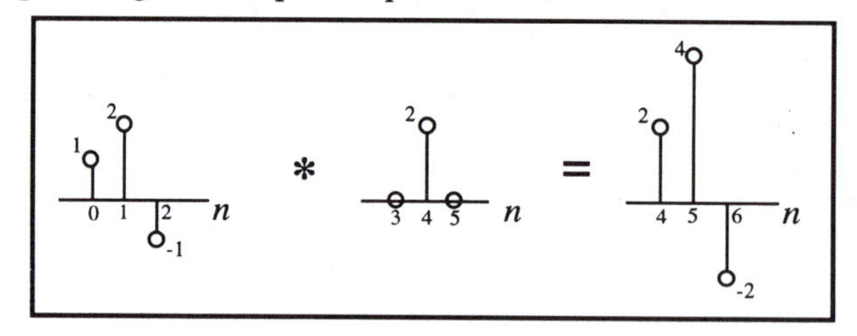

11.3 Convolution Through Signal Decomposition

If both $x[n]$ and $h[n]$ are relatively short in length, the signal decomposition method provides an extremely fast and easy way to calculate $y[n]$, while completely avoiding the entire flip/shift process. The idea is to break up $h[n]$ into individual impulses (pick the shorter signal to be $h[n]$). Each impulse then produces a copy of $x[n]$, appropriately scaled and shifted as described in Section 11.2. Since $h[n]$ is just a sum of impulses, $y[n]$ will simply be the sum of the scaled and shifted copies of $x[n]$. For further clarity and comparison purposes, we will illustrate this process using the same $x[n]$ and $h[n]$ from the Example in Section 11.1.

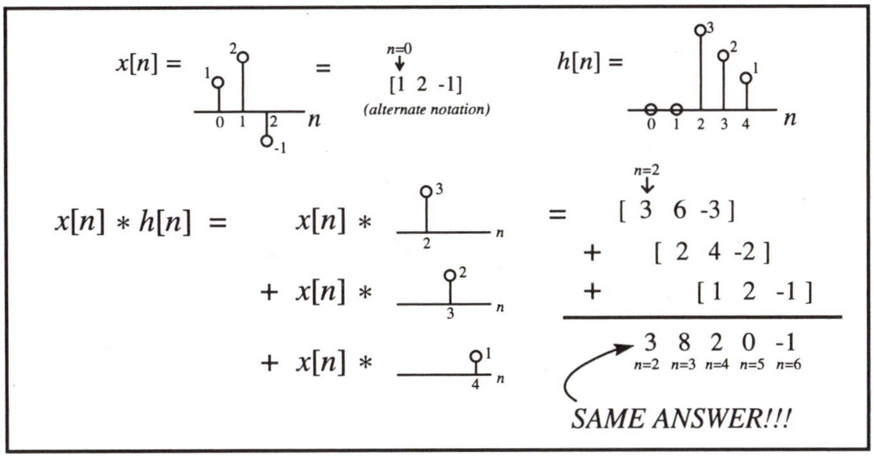

$x[n] =$ (stem plot: values 1, 2, -1 at n = 0, 1, 2) $=$ $\overset{n=0}{\downarrow}$ [1 2 -1] *(alternate notation)*

$h[n] =$ (stem plot: values 0, 0, 3, 2, 1 at n = 0, 1, 2, 3, 4)

$x[n] * h[n] = x[n] * 3\delta$ at $n=2$ $+ \; x[n] * 2\delta$ at $n=3$ $+ \; x[n] * 1\delta$ at $n=4$

$=$ $\overset{n=2}{\downarrow}$ [3 6 -3]
$+$ [2 4 -2]
$+$ [1 2 -1]

3 8 2 0 -1
n=2 n=3 n=4 n=5 n=6

SAME ANSWER!!!

11.4 Convolution of Infinite Length Signals

When either $x[n]$ or $h[n]$ are infinite in length, it is often necessary to employ purely algebraic techniques to find the output signal $y[n]$.

Example

$$x[n] = \left(\frac{1}{2}\right)^n u[n]$$

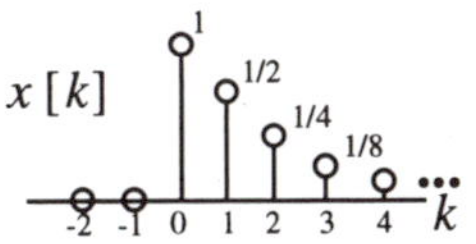

$x[k]$ (stem plot: values 1, 1/2, 1/4, 1/8, ... at k = 0, 1, 2, 3, 4; zero at k = -2, -1)

$$h[n] = u[n]$$

$h[n-k]$ (stem plot: value 1 for $k \le n$, ... ; zero at $n+1$, $n+2$; axis labels $n-4$, $n-2$, n, $n+2$)

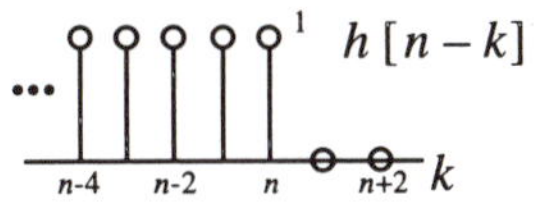

For $n \geq 0$,

$$y[n] = \sum_{k=0}^{n} \left(\frac{1}{2}\right)^k = \frac{1-\left(\frac{1}{2}\right)^{n+1}}{1-\frac{1}{2}}$$

$$\Rightarrow \quad \boxed{y[n] = \left(2-\left(\frac{1}{2}\right)^n\right)u[n]}$$

Some Useful Formulas:

Infinite Geometric Series

$$a_1 \sum_{n=0}^{\infty} r^n = a_1 + a_1 r + a_1 r^2 + a_1 r^3 + \ldots = \frac{a_1}{1-r} \qquad |r| < 1$$

Finite Geometric Series

$$a_1 \sum_{n=0}^{N} r^n = \underbrace{a_1 + a_1 r + a_1 r^2 + \ldots + a_1 r^N}_{\text{N+1 terms}} = \frac{a_1(1-r^{N+1})}{1-r}$$

See the Appendix for a more complete discussion of sequences and series.

11.5 Useful Checks

For $y[n] = x[n] * h[n]$, the length of the output signal $y[n]$ is generally equal to $length(x[n]) + length(h[n]) - 1$. The output signal will begin at $start(x[n])+start(h[n])$ and finish at $end(x[n])+end(h[n])$. Note that with some lucky cancellations the output signal might actually be shorter than expected, but it will never be longer. Also note that a symmetric signal convolved with a symmetric signal will always yield a symmetric output signal.

11.6 Convolution Intuition

By now you should start to have a feel for what convolution does to a signal. Imagine you are given an $x[n]$. Look at the example $h[n]$'s shown below and try to determine what effect the convolution process will have on that signal. Just imagine flipping and sliding $h[n]$ through the input – what's happening? If you can look at two signals and roughly sketch what the result of convolving them would be, then you're on your way to signal processing nirvana.

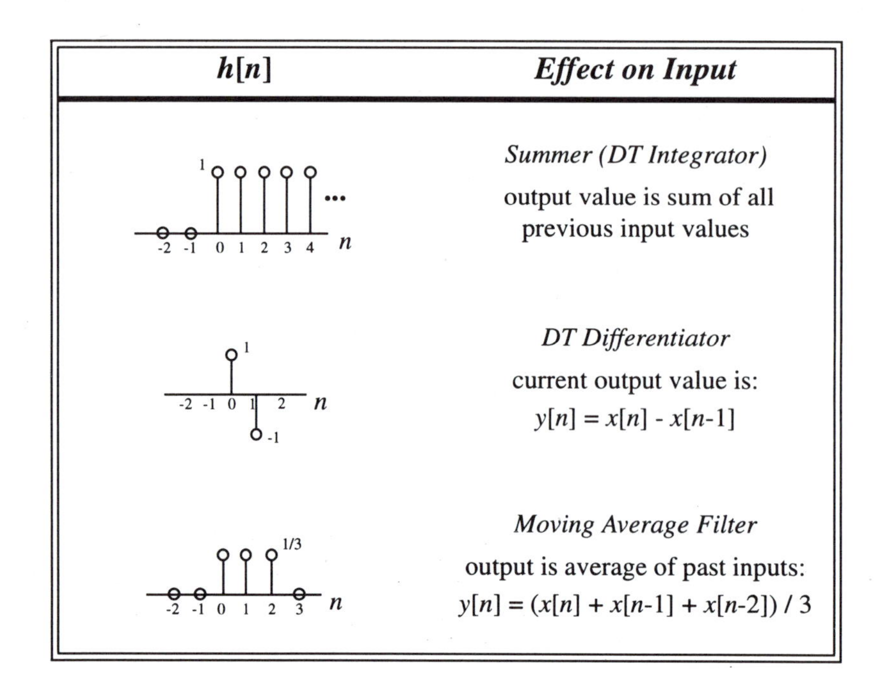

CHAPTER 12 Continuous-Time Convolution

Overview

This chapter describes the mechanics of continuous-time convolution. It offers a step-by-step procedure derived directly from the convolution integral. There are also hints and shortcuts provided for developing a more intuitive feel for the convolution process. The chapter concludes with a discussion of matched filters, one of the many practical uses of convolution.

12.1 Graphical Flip/Shift Method

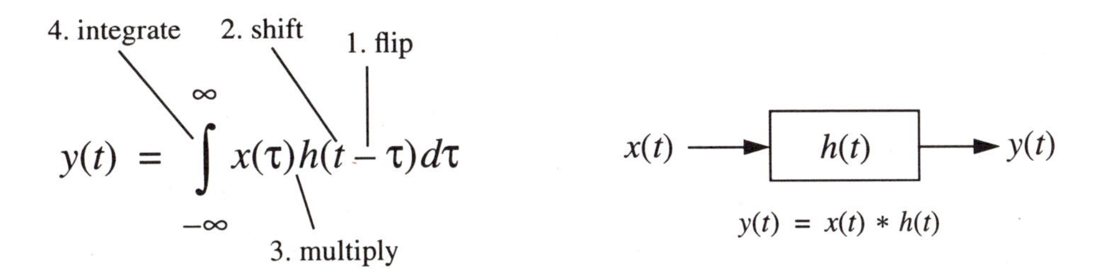

Looking at the convolution equation above, we can arrive at the following step-by-step procedure:

1. Choose one signal to be $x(t)$, the other is then $h(t)$; draw them both on the τ axis.
2. *FLIP* $h(\tau)$ about $\tau=0$ and *SHIFT* signal to the right by t.
3. Identify the different regions of integration (look for breakpoints in the signals).
4. *MULTIPLY* $x(t)$ by flipped/shifted version of $h(t)$ and *INTEGRATE* using correct limits on integral.
5. Step 4 produces the equation of $y(t)$ over the specified region.
6. Repeat step 4 for all possible regions of interest.

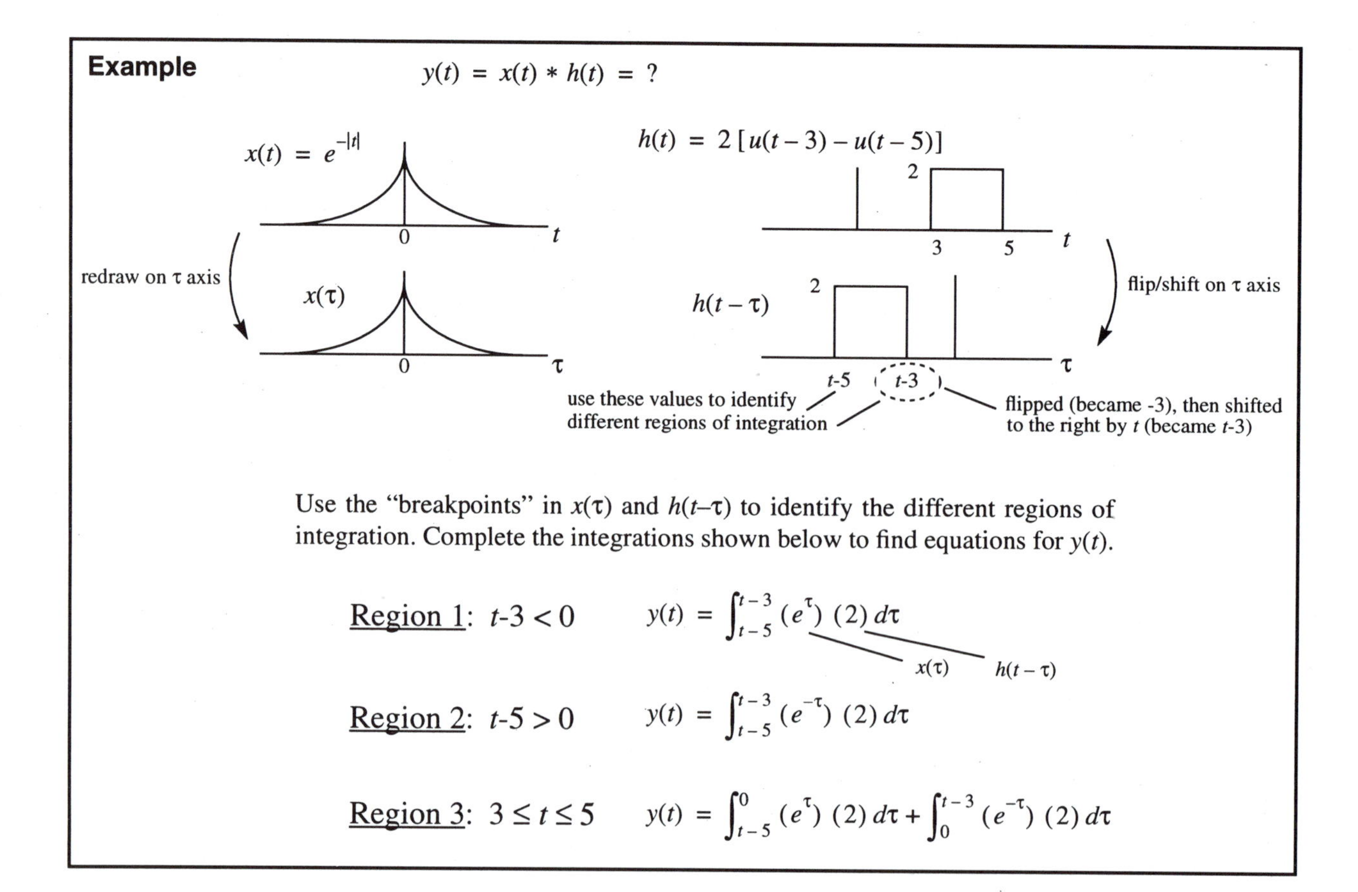

Here's another sample problem:

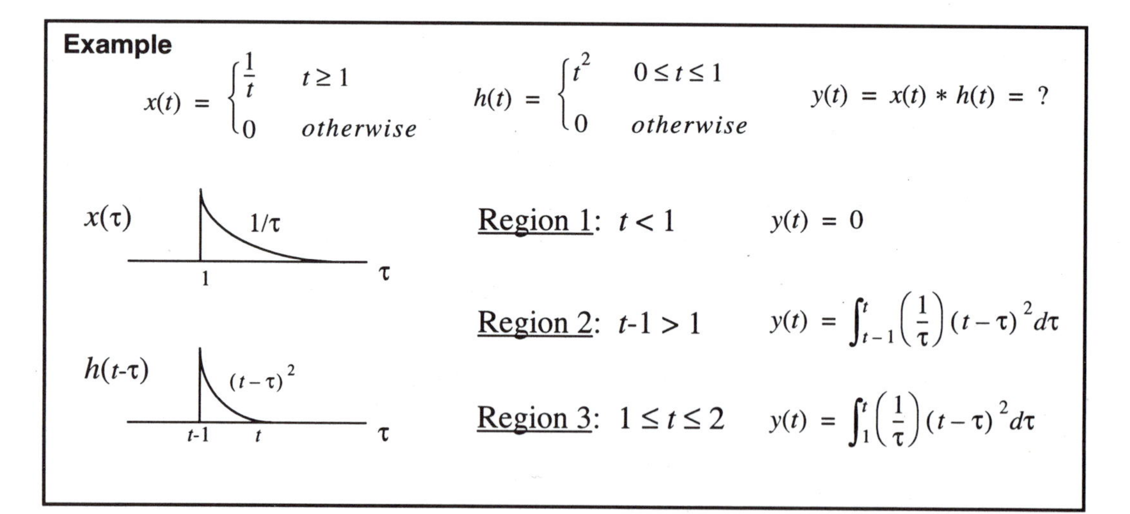

12.2 Convolution with Impulses

Convolution with an impulse simply produces a copy of the signal scaled and shifted by the size and location of the impulse. That's all! No flipping, no shifting, no integration, just write the answer down! Similarly, when convolving with a doublet, just write down the derivative of the original signal (appropriately shifted).

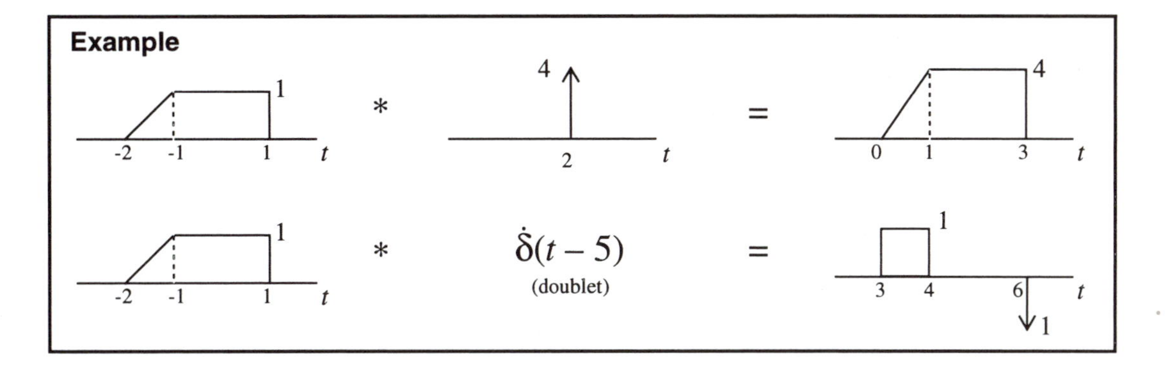

12.3 Convolution by Inspection

Aside from convolving with impulses or doublets, there are other times when it is actually quite easy to just write down the result of a convolution. This method relies on a mental picture of flipping and shifting and works particularly well for signals with "box-like" shapes. Start by visualizing flipping and sliding one of the signals (choose the simpler looking one) through the other signal. Stop at appropriate breakpoints (e.g. step transitions in the signal), multiply the signals, and then compute the area of the product. Note, this is not, repeat NOT, the same thing as looking for the area *of the overlap*. Plot the result as a single point on the axes for the answer. Repeat this procedure for all appropriate transition points and simply connect the dots.

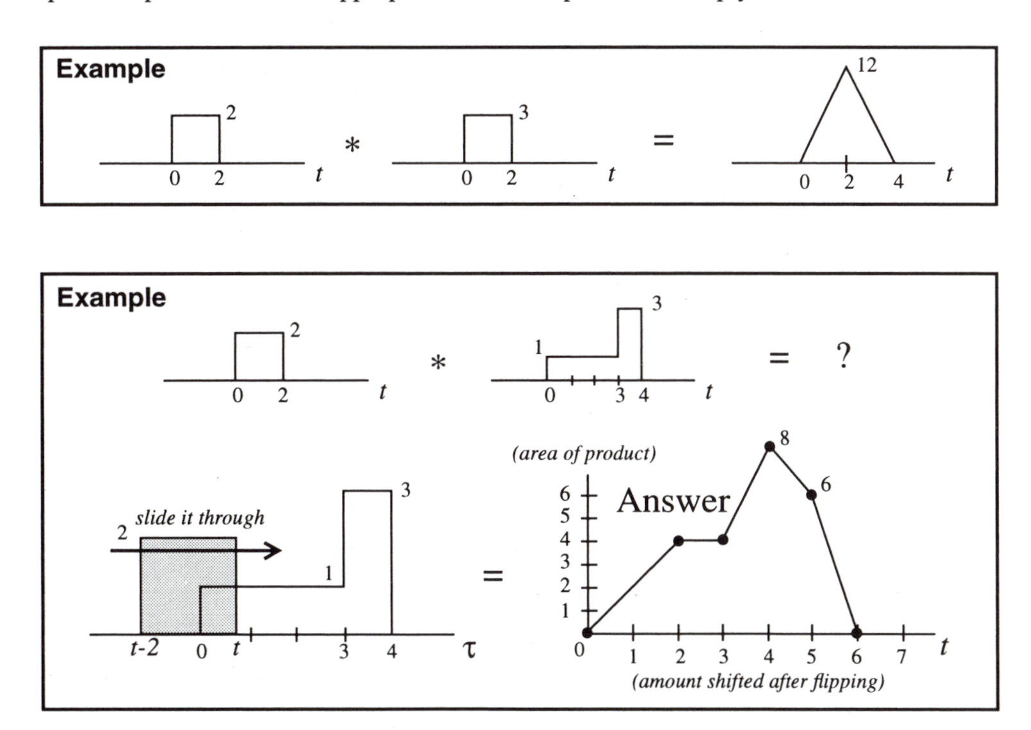

12.4 Useful Checks

The result of convolving continuous-time signal $A(t)$ with signal $B(t)$ generally produces a signal $C(t)$ that is the length of $A(t)$ plus the length of $B(t)$. Furthermore, signal $C(t)$ will start at $start(A(t)) + start(B(t))$ and end at $end(A(t)) + end(B(t))$. This is a simple way to partially check your work when doing convolutions. Another useful tip is that a symmetric signal convolved with a symmetric signal always produces a symmetric signal.

12.5 Matched Filters

There are often situations when you would like to "pick out" or identify the location or existence of a particular shape or object buried within another signal. Examples include trying to identify the transmission of messages over a noisy or secret communication channel, identifying an enemy tank from an image taken from the nose of a cruise missile, or in this case, pinpointing the location in time of each heartbeat during an electrocardiogram recording in order to help identify irregular beat patterns known as arrithymias. The general idea is to convolve your signal with a template of the signal you are searching for. This template is known as the *matched filter.* Strictly speaking, the matched filter should actually be the time-reversed version of the signal you are searching for, since during convolution you will flip it back again. During the convolution, if the matched filter overlaps with something of similar size and shape, the output of the convolution will be very large (remember, when convolving, multiply the signals and then integrate).

The purpose of this example is to try and identify the location in time of the triangle-shaped QRS complex that is part of each heartbeat. The convolution produces a sharp peak at the time of the QRS even in the presence of an incredible amount of noise (baseline drifts, extraneous skeletal muscle contractions, etc.). Now, a simple thresholding scheme can be used to pick off the exact location of each heartbeat.

Clean ECG data

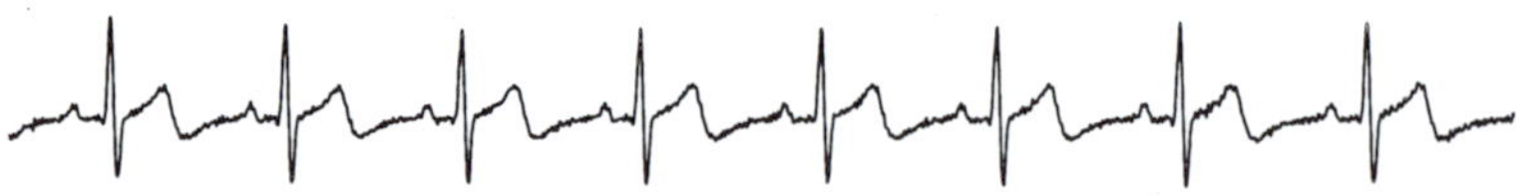

ECG data with Added Sinusoidal and Random Noise

Closeup of Matched Filter

Result of Convolution with Matched Filter

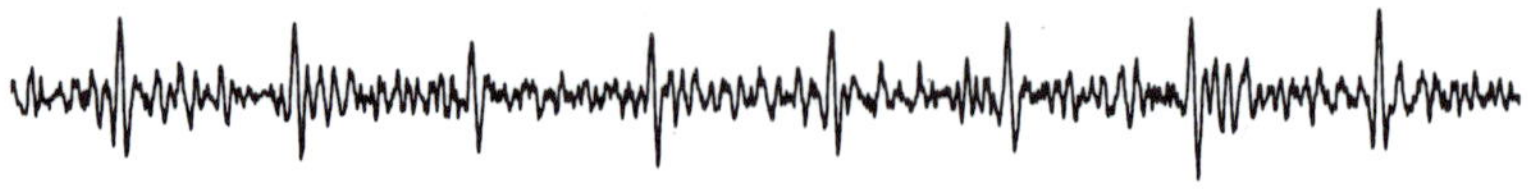

CHAPTER 13 Deconvolution

Overview

Deconvolution is the process of recovering the input signal when given the output of a system. It is the inverse of convolution and has many practical applications, especially in image processing. Two potential pitfalls when doing deconvolution, excess noise and instability of the inverse filters, are briefly discussed.

13.1 What is Deconvolution?

Deconvolution is the process of "undoing" a convolution operation. That is, given the output of a system and its impulse response $h(t)$ or transfer function $H(s)$, try to determine what the input to the system must have been to produce that output.

13.2 How to do Deconvolution

Here is a graphical representation of how deconvolution would work in the time domain. Here, $\hat{h}(t)$ is known as the inverse or deconvolving filter.

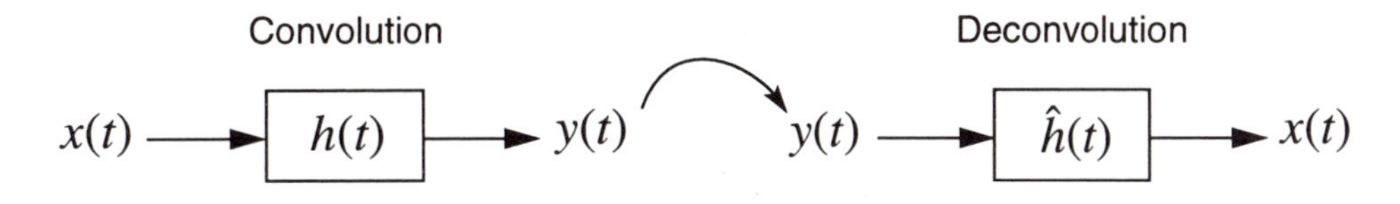

So how can we find this magical $\hat{h}(t)$? Well, from the above block diagram, we see that it must be the case that $h(t) * \hat{h}(t) = \delta(t)$ (since $x(t) * \delta(t) = x(t)$). So, in the frequency domain, this means that $H(s)\hat{H}(s) = 1$. Thus, we can find $\hat{h}(t)$ by taking the inverse transform of $1/H(s)$. Repeating once more, the transform of $h(t)$ is the reciprocal of the transform of $\hat{h}(t)$.

In discrete time, we have the similar results of $h[n] * \hat{h}[n] = \delta[n]$ or $H(z)\hat{H}(z) = 1$.

13.3 Why is it Useful?

Deconvolution has many practical uses, especially in image processing. Note that working with images involves doing two-dimensional convolutions, but the concept is essentially the same. For example, astronomers who have a good model of atmospheric effects (i.e. know its 2-D impulse response or "point spread function") can perform a deconvolution back on Earth to sharpen up their images.

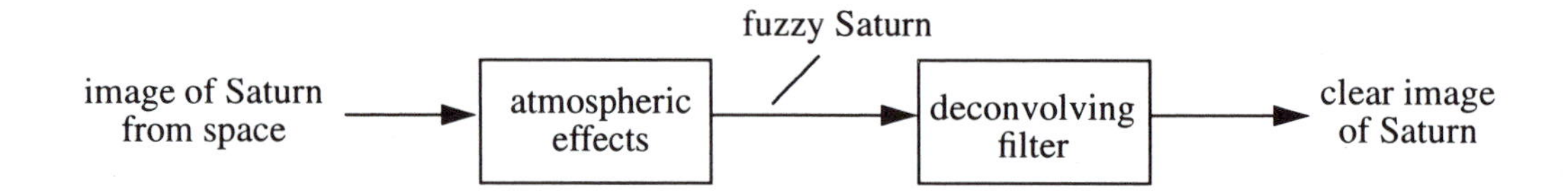

Other uses of deconvolution include removing the linear blurring that occurs when taking a picture of a speeding race car and trying to recover a secret communication signal after it has been deliberately garbled by a convolution. Note that in all of these cases, complete reconstruction of the original signal requires an exact knowledge of the original system function $H(s)$. However, in the real world, that situation is rarely the case. There is a heuristic technique known as "blind deconvolution" to help overcome this problem. This method, as its name implies, is a way of doing the deconvolution without knowing the exact equation of the original impulse response. To learn more about blind deconvolution and other image processing techniques, refer to *Two-Dimensional Signal and Image Processing* by Lim (1990). If you become a deconvolution expert, the CIA will beat down your door to hire you.

13.4 Potential Pitfalls

Noise

Let's return to the Saturn example, but this time include the very real possibility of additive noise:

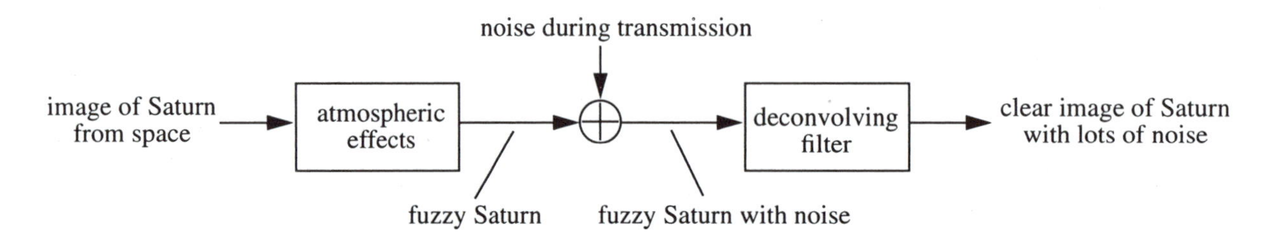

If your original system function $H(s)$ performed some sort of smoothing or blurring on your data (as is the case with the majority of systems), then if deconvolution is to work, it must "unsmooth" or sharpen the data. If there is noise present, the sharp transitions associated with the additive noise will by amplified by deconvolution, making your image look a lot worse. The moral is to think carefully about noise when doing deconvolutions.

Stability

Since $\hat{H}(s)$ is the reciprocal of $H(s)$, it should be obvious that the zeros of $H(s)$ become the poles of $\hat{H}(s)$ (and vice versa). Thus if $H(s)$ has a zero in the right-half plane, the inverse filter $\hat{h}(t)$ becomes unstable. Or if you want a stable version of $\hat{h}(t)$, then it cannot be causal – which doesn't make much sense in the real world. Handling deconvolutions on systems that have zeros in the right-half plane, or outside the unit circle in the case of discrete time, requires more sophisticated techniques. Also remember that in discrete time, as long as the data can be processed off-line (i.e. not in real-time), causality is not much of a concern.

CHAPTER 14 Causality and Stability

Overview

This chapter covers two important LTI system properties: causality and stability. After defining these terms, the appropriate criteria for a system to be causal or stable are presented and explained.

14.1 What is Causality?

Causality is a property of a system. An LTI system is *causal* if the output is dependent only on the current and/or past values of the input signal; loosely speaking, the input causes the output. In other words, in a causal system the current output value is not dependent on future values of the input; the system cannot anticipate what is coming up next and alter its output accordingly. Intuitively, it should seem clear that any real-world electrical or mechanical system is inherently causal. For example, the output of a circuit will change as soon as or slightly after the input is applied, but never before; or a car will behave exactly the same at time t, regardless of whether you plan to slam the brakes or floor the accelerator at time $t+1$.

It sounds like practically every system is causal, so what's the big fuss about? Causality is more of an issue in discrete-time systems. For example, in image processing, modification of the current pixel value may depend on both future and prior pixel values. This is not a problem since the data has already been collected and there is no real notion of "time." As for continuous time, you've probably noticed that some books define the Laplace transform from $-\infty < t < \infty$ (bilateral) while others use $0 \le t < \infty$ (unilateral). Those using the unilateral transform are assuming that all systems are causal (no need for negative time). The bilateral transform on the other hand is more general and has some nice mathematical properties.

14.2 Condition for Causality

A system is causal if and only if its impulse response is zero for t or n less than zero. Similarly, *anti-causal* systems have an impulse response that is zero for t or n greater than zero.

$$\text{Causal System} \Leftrightarrow \begin{array}{ll} h(t) = 0 & t < 0 \\ h[n] = 0 & n < 0 \end{array}$$

This requirement for causality is better understood if viewed in the context of convolution. If $h(t) \neq 0$ for all $t < 0$, then when flipping and shifting during convolution, the current output value would depend on future inputs, which goes against the notion of causality. This process is illustrated in the following example:

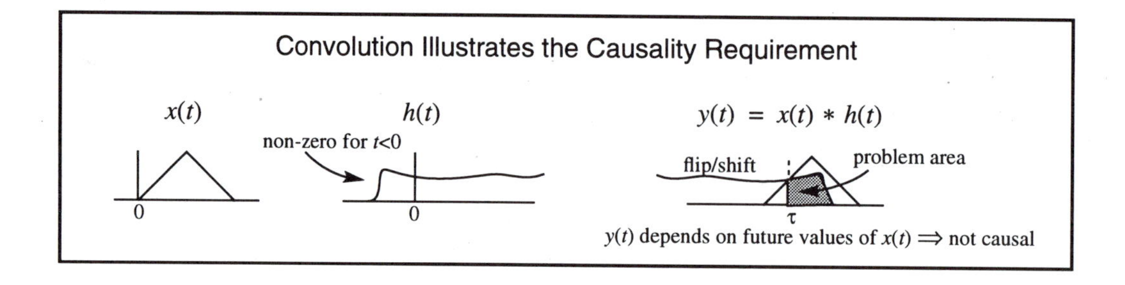

14.3 What is a Stable System?

What does it mean for a system to be stable? First, let's present an intuitive description. Assume there is a system sitting in front of you with currently no input or output signal present. Suddenly you turn on a finite input signal. This input can be mechanical, electrical, acoustic, or even in the form of light energy. In any case, the system starts to produce an output. If the input stops or remains finite and the output signal continues to grow larger and larger until reaching infinite proportions, this system is said to be *unstable*. If the output remains at a constant size forever even though the input signal has long since stopped, the system is said to be *marginally stable*. And finally, if the input stops and the output slowly decays away, or else the input continues and the output signal behaves itself (doesn't blow up) then the system is said to be *stable*. Here is a visual description of the three types of stability:

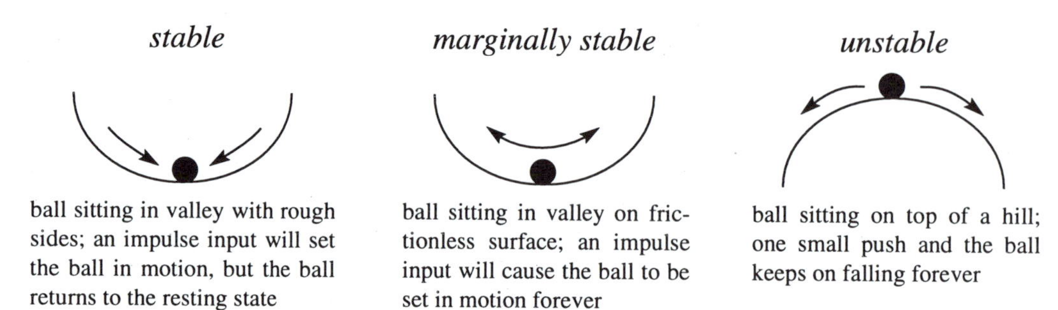

ball sitting in valley with rough sides; an impulse input will set the ball in motion, but the ball returns to the resting state

ball sitting in valley on frictionless surface; an impulse input will cause the ball to be set in motion forever

ball sitting on top of a hill; one small push and the ball keeps on falling forever

The term "stable" normally refers to stability in the BIBO sense, meaning bounded-input-bounded-output. In other words, an LTI system is stable if all bounded inputs (i.e. finite amplitude) produce a bounded output signal (again, finite amplitude). Don't confuse "very big" with infinite. For example, an amplifier with a very high gain is not an unstable system. A finite sized input will still produce a finite sized output, albeit very large.

> An LTI system is BIBO stable if all bounded inputs (no matter how large) produce bounded outputs (i.e. never go to infinity).

14.4 Conditions for Stability

There are many ways to insure that a system is BIBO stable. Three of the most common methods are discussed below.

(1) Impulse Response

A system is BIBO stable if and only if its impulse response is absolutely summable/integrable. This is easy to understand if you visualize the convolution process using a typical bounded input, like a unit step function.

$$\text{BIBO Stability} \Leftrightarrow \int_{-\infty}^{\infty} |h(\tau)|\, d\tau < \infty \quad \text{or} \quad \sum_{k=-\infty}^{\infty} |h[k]| < \infty$$

(2) ROC of System Function

A system is BIBO stable if and only if the region of convergence (ROC) of its system function includes the $j\omega$-axis (continuous-time systems) or the unit circle (discrete-time systems).

Recall that for causal systems, or those with right-sided impulse responses, the ROC is to the right of the right-most pole (CT) or outside the outermost pole (DT). For causal systems, the stability requirement then translates into saying that all the system poles are in the left-half plane $(Re\{s\} < 0)$ for CT systems or inside of the unit circle $(|z| < 1)$ for DT systems. A system that has poles directly on the $j\omega$-axis or the unit circle is said to be marginally stable. It is important to note that the locations of zeros have no effect on stability.

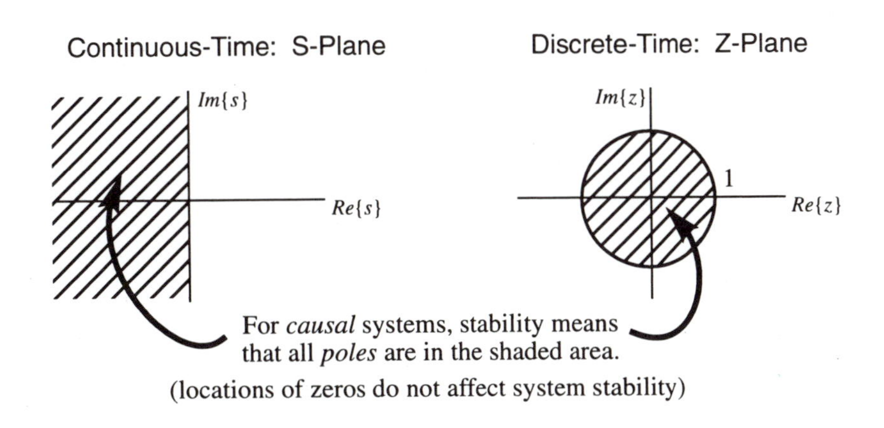

(3) Characteristic Equation

As shown in the previous diagram, stability for continuous-time causal systems means that all poles are in the left-half plane. This implies that the roots of the characteristic equation (denominator of the system function $H(s)$) are all negative, or at least have negative real parts. The Routh-Hurwitz criterion is a mathematical procedure for determining whether or not this is true based only upon the coefficients of this polynomial.

For all order systems a necessary but not sufficient condition for full stability (not marginal) is that all characteristic polynomial coefficients are present and are of the same sign. The necessary and sufficient criteria for a first, second, or third order system to be stable are shown in the table below.

System Order	Characteristic Equation	Stability Criteria
1	$s + a = 0$	$a > 0$
2	$s^2 + as + b = 0$	$a > 0, b > 0$
3	$s^3 + as^2 + bs + c = 0$	$a, b, c > 0$ and $ab > c$

For higher order systems, all coefficients positive is a necessary, but not sufficient condition.

To insure stability for discrete-time causal systems, we must verify that the magnitudes of all pole locations are less than one (i.e. inside the unit circle). Jury's test (the DT equivalent of Routh-Hurwitz) may be applied, but it is not a simple procedure. Instead, solve for the roots by hand or use a computer package like MATLAB.

CHAPTER 15 Feedback

Overview

Feedback is the process of using the output of a system to continually alter or update its input. This type of connection results in a modification of the original transfer function for the purpose of increasing stability, removing nonlinearities, or simply modifying the dynamic response. This chapter explains how to quickly analyze feedback block diagrams, warns against some common pitfalls regarding loading effects, and illustrates some uses and advantages of feedback with a few practical examples.

15.1 What is Feedback?

In previous chapters we have seen block diagrams where systems are connected together either in series or parallel. Feedback is a special type of system interconnection in which the output of the system is "fed back" to the input, possibly through the addition of other systems. Feedback is the process of using the output of a system to continually alter or update its input. For example, let's say you're trying to balance a broomstick on the palm of your hand. You take the output of the system (the angle of the stick and its velocity) and accordingly alter the input (your hand position) to keep the broomstick balanced. Feedback systems are also often called "closed-loop" systems. An example of an "open-loop" approach would be charting an airplane course across the country, predicting all wind speeds and directions in advance based upon your system model of the atmosphere, flying with your eyes closed, and hoping you land in the right place. As you can imagine, this is rather foolish. Rather, if you continually checked your position against your predicted flight path, you could make small adjustments, thus creating a closed-loop situation. Another example is illustrated below:

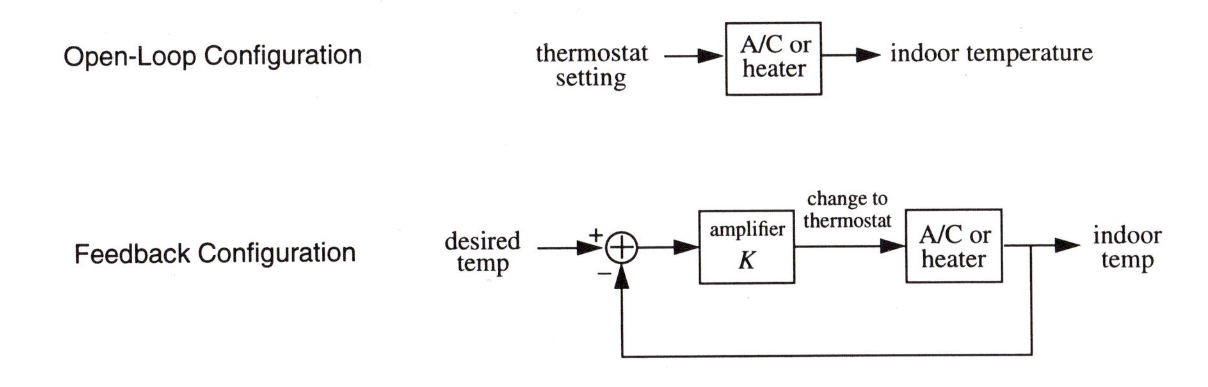

By incorporating feedback, the overall transfer function of the original system is altered. This process can be used to stabilize unstable systems, improve disturbance rejection, alter the dynamic or transient response characteristics, and even remove system nonlinearities. Feedback is the basis of most practical systems.

15.2 Positive versus Negative Feedback

Notice the minus sign in the feedback loop in the block diagram from Section 15.1. This minus sign implements what is known as *negative feedback*. The input to the system is adjusted opposite to the direction the output is moving. For example, if the room is too hot, the feedback loop tries to make it colder; if the room is too cold, the system tries to make it warmer. Think about what would happen if the minus sign was changed to a plus, which is known as *positive feedback*. Things would blow up! Very shortly you would begin to either freeze or sweat. Another example of positive feedback is the following innocuous looking system:

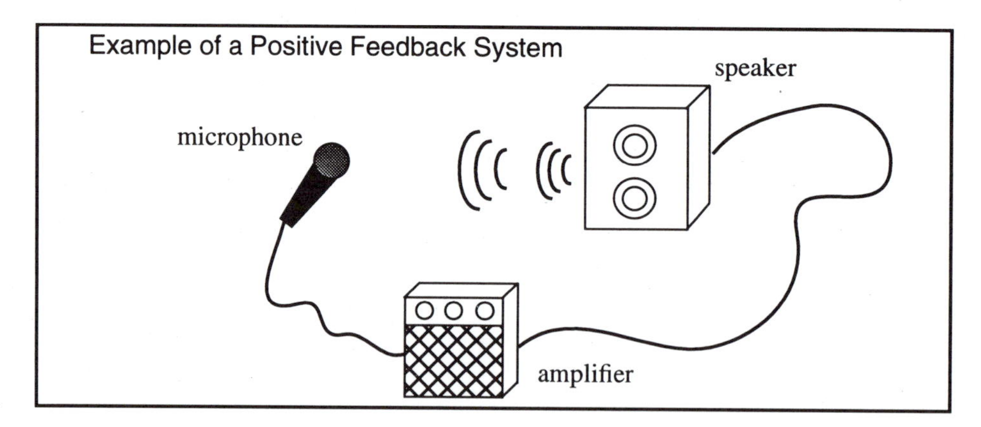

Have you ever heard that annoyingly loud whine that occasionally comes out of a microphone/amplifier/speaker system? Why does it do that? Well, little noises near the microphone get picked up by the mike, amplified, and then played through the speaker, which creates a louder sound – which is picked up by the mike, amplified, and played through the speaker, and so on. In a few tenths of a second a mere breath blown on the mike gets turned into an annoyingly loud screech by the speaker (the amplifier or speaker is saturated). Keeping the microphone sufficiently far enough away from the speaker generally solves the problem. As you can see, things hooked up in positive feedback configurations are generally thought to be unstable. The stability of feedback systems will be discussed in further detail in Section 15.8.

15.3 Black's Formula

Black's formula is a method for deriving the equivalent transfer functions of systems that contain feedback loops. But first, let's start off with a simple example and show how to derive $H(s)$ the long way:

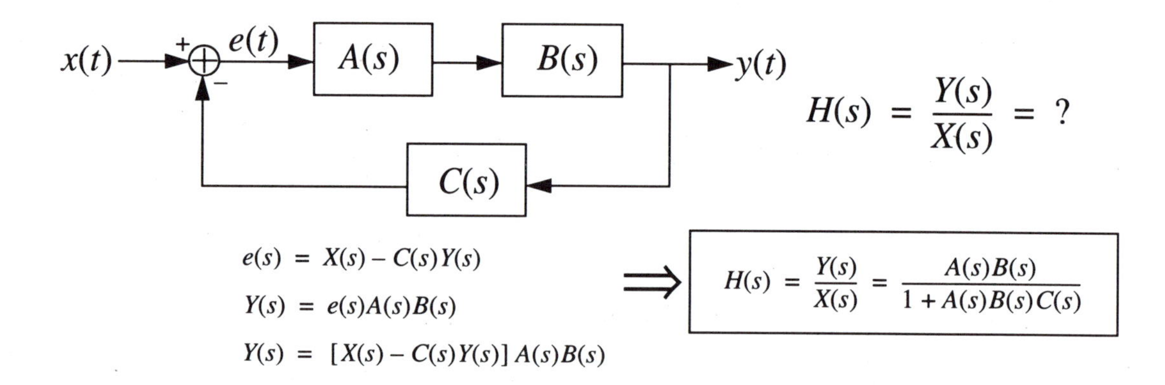

$$H(s) = \frac{Y(s)}{X(s)} = ?$$

$$e(s) = X(s) - C(s)Y(s)$$
$$Y(s) = e(s)A(s)B(s)$$
$$Y(s) = [X(s) - C(s)Y(s)]\,A(s)B(s)$$

$$\Rightarrow \quad H(s) = \frac{Y(s)}{X(s)} = \frac{A(s)B(s)}{1 + A(s)B(s)C(s)}$$

This means that the original feedback loop can be replaced with a single box:

$$x(t) \longrightarrow \boxed{\frac{A(s)B(s)}{1 + A(s)B(s)C(s)}} \longrightarrow y(t)$$

Black's formula is a way of achieving this same result directly, without any equation writing. It can be directly applied to general feedback configurations as follows:

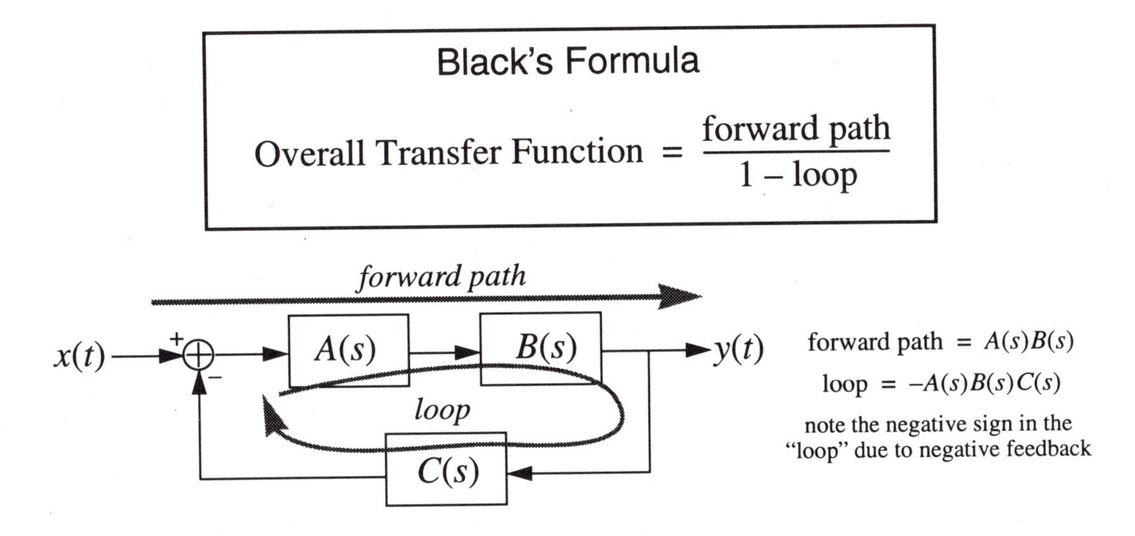

More complicated systems may have multiple feedback loops in them. Break down such beasts by replacing embedded single feedback loops with a single box using Black's formula. Repeat this process as many times as necessary and soon seemingly immense systems become quite manageable.

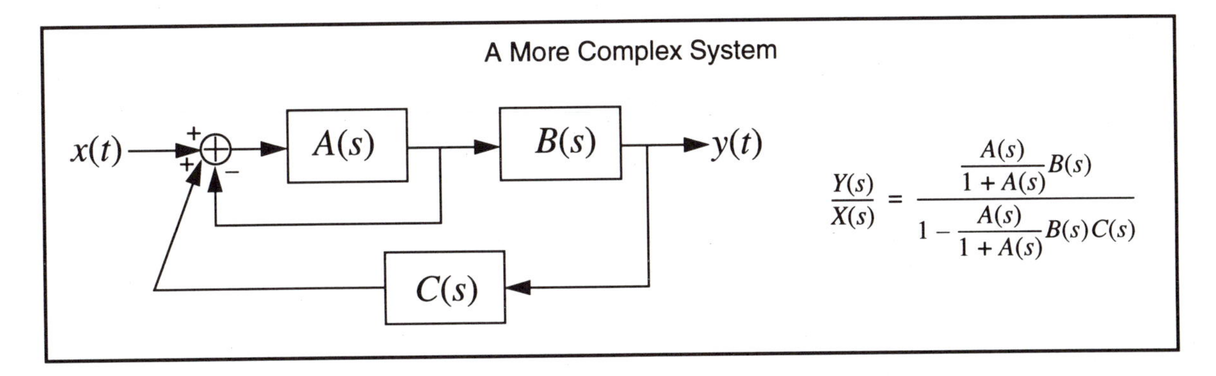

Should you ever forget Black's formula or just plain become confused, you can always resort to the long, but straightforward method of first writing down all possible equations relating $x(t)$, $y(t)$, and as many intermediate variables like $e(t)$ that are needed. Then, repeatedly reduce/substitute equations together until a relationship between x and y is found (just like the method at the start of this section).

15.4 Loading Effects

Although it seems straightforward to reduce large systems (several blocks) into a single block, we cannot afford to be so carefree without considering *loading effects*. In other words, the characteristics of a particular system may be affected by what precedes or follows it. This is generally not a good feature and is the sign of a poor design. Ideally, box #1 should behave the same regardless of whether box #2 is there or not. Adding the second box should not "load down" or suck current out of the first box. This can be achieved one of two ways: (1) the second box has a very high input impedance – meaning it lets no current in, or (2) the first box has zero output impedance – meaning it can supply an infinite amount of current if necessary. If either of these two conditions is satisfied, then we are free to combine boxes together in the traditional series fashion.

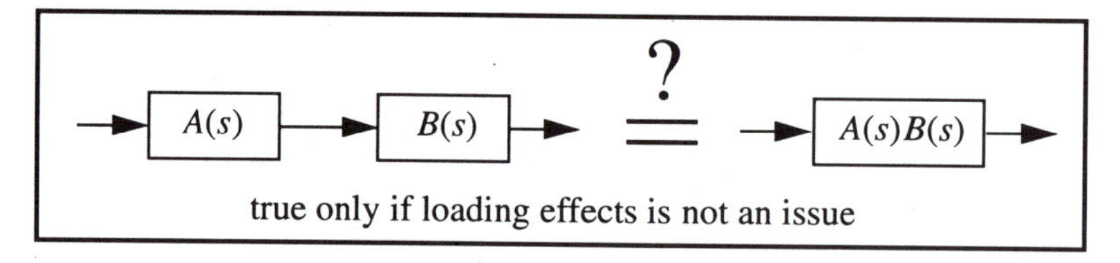

A concrete example of the dangers of carelessly combining boxes together:

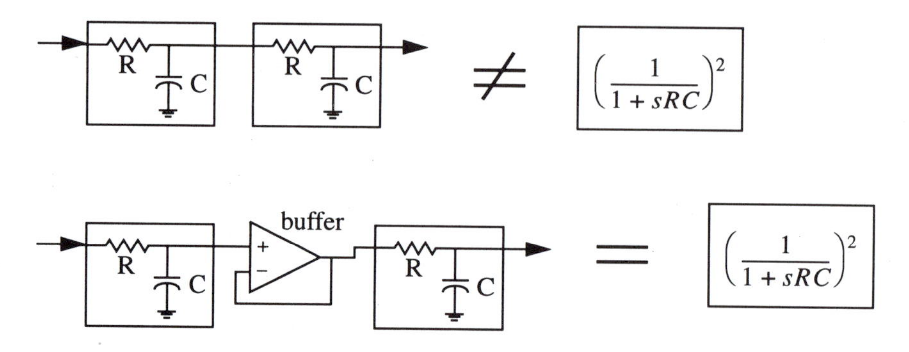

15.5 Using Feedback to Invert a System

Feedback can also be used to produce the inverse of a system. Watch what happens when a system $B(s)$ is put in the middle of a feedback loop, with the loop gain K turned up quite high:

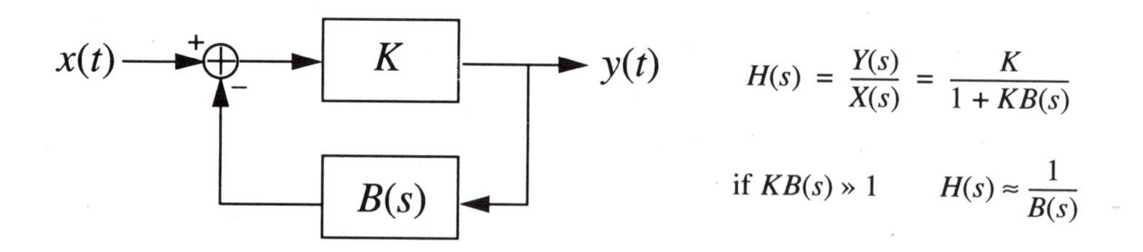

Using Black's formula, and letting $K \to \infty$, we see that the resulting overall transfer function is now $1/B(s)$, or the inverse of the original system. So remember, one way to invert a system is to put the system in a feedback loop – simple, yet powerful. Note that to insure $1/B(s)$ is stable, all the zeros of $B(s)$ must be in the left-half plane (assuming it is a causal system).

15.6 Accounting for System Fluctuations

Feedback can also be used to remove the effects of system uncertainties. For example, let's say you've built a great stereo power amplifier that's supposed to have a gain of one. However, for some reason the gain varies between 0.5 and 10 depending on the outside temperature. This problem can be fixed by using feedback combined with a high gain preamplifier as follows:

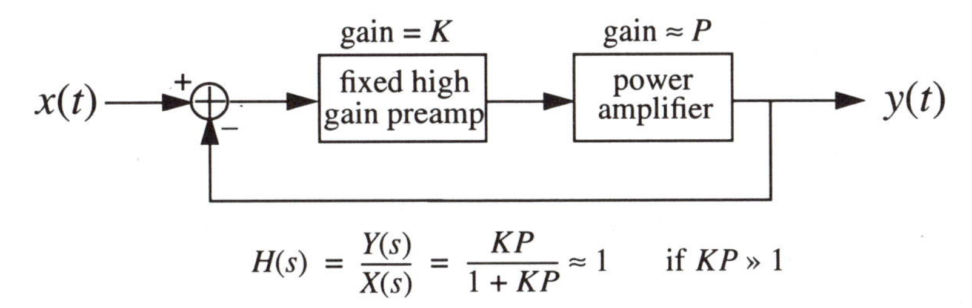

15.7 Removing System Nonlinearities

Let's continue with the previous example and assume that not only is the gain of the power amplifier a bit unpredictable, but also that its transfer function is nonlinear. This is typical of many power amplifier stages due to the turn-on voltages necessary for diodes and transistors. Not to fear however, a very quick analysis of the following block diagram using Black's formula should convince you that feedback can be used to eliminate or at least clean up this problem. See Example 5.3-3 of *Circuits, Signals, and Systems* by Siebert (1986) for a more detailed explanation.

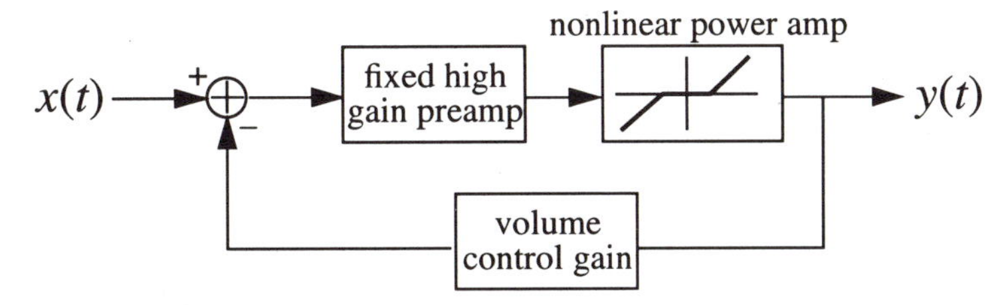

15.8 Using Feedback to Stabilize Systems

The following causal system is clearly unstable:

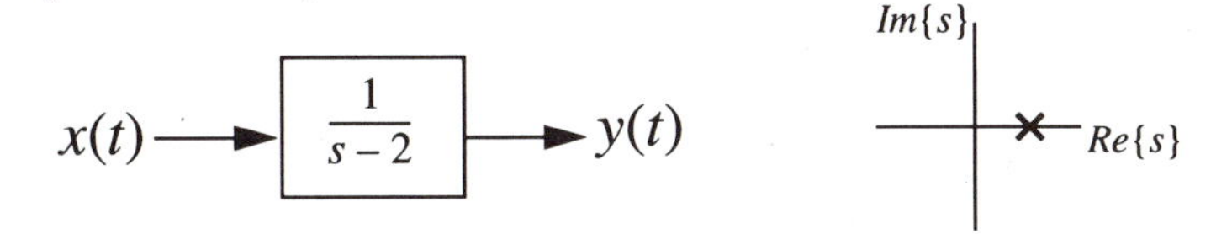

Nevertheless, it is possible to stabilize it by putting it in a feedback loop as follows:

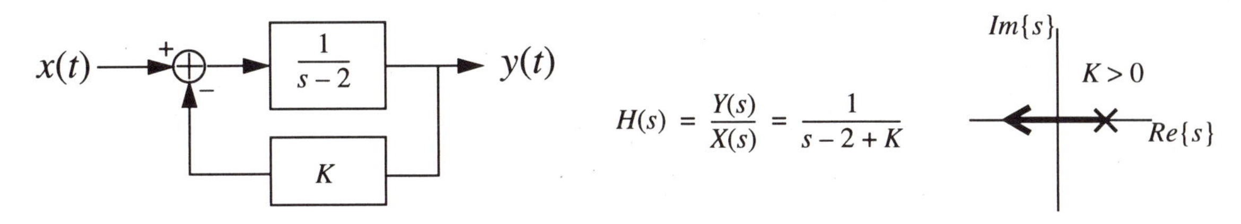

For what values of K is this system stable? Recall that we want all system poles in the left half plane. As we increase the feedback gain K from zero, the pole begins to move to the left. The manner in which the system poles change location as a function of the feedback gain is known as the *root-locus*. These pole movements are easily predicted according to a set of well-defined rules. Knowing the pole locations can supply information not only about stability, but also about factors such as step response overshoot, resonant frequencies, etc.

The *Nyquist stability criterion* is another way to assess stability of closed loop systems. Its advantage is that it does not require explicit knowledge of the system function; in fact, the system function doesn't even have to be rational. Yet another method of verifying closed-loop stability is to determine what is known as the *gain and phase margins* of the Bode plot of the closed-loop transfer function. For a more thorough discussion of root-locus, Nyquist stability criterion, gain/phase margins, or feedback in general, see Chapter 11 of *Signals and Systems* by Oppenheim et al (1983).

15.9 A Sample Problem

For what values of K is the following causal closed-loop discrete-time system stable?

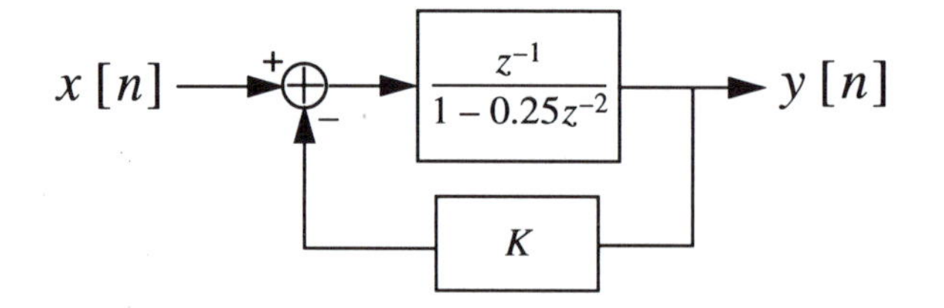

Solution:

$$\tilde{H}(z) = \frac{\tilde{Y}(z)}{\tilde{X}(z)} = \frac{z^{-1}}{1-0.25z^{-2}+Kz^{-1}} = \frac{z}{z^2+Kz-0.25}$$

$$\text{poles using quadratic formula} = \frac{-b \pm \sqrt{b^2-4ac}}{2a} = \frac{-K \pm \sqrt{K^2+1}}{2}$$

$$\text{want } \left|\frac{-K \pm \sqrt{K^2+1}}{2}\right| < 1 \text{ for DT stability}$$

solving inequalities involving radicals is a tricky process; let's assume an equality in order to first obtain the boundaries of the solution:

$$\frac{-K \pm \sqrt{K^2+1}}{2} = 1 \quad \text{or} \quad \frac{-K \pm \sqrt{K^2+1}}{2} = -1$$

$$K^2+1 = K^2+4K+4 \qquad K^2+1 = K^2-4K+4$$

$$K = -\frac{3}{4} \qquad K = \frac{3}{4}$$

Is the solution $|K| < 3/4$ or is it $K > 3/4$, $K < -3/4$?

$K = 0$ solves the inequality, so the solution must be:

$$|K| < \frac{3}{4}$$

CHAPTER 16 The Fourier Transform

Overview

The Fourier transform is one of the most commonly used techniques in signal processing. It is simply a mathematical transformation that changes a signal from a time-domain representation to a frequency-domain representation, allowing one to analyze the "frequency content" of a signal. This chapter first provides an intuitive overview of the transform and then proceeds to illustrate many of its mathematical properties that aid in its computation and understanding. Try not to get lost in the math; processing a signal in the frequency domain has numerous practical advantages, as will become clear in later chapters.

16.1 What is the Fourier Transform?

The Fourier Transform is one of the most commonly used techniques in signal processing. This formula is a mathematical transformation that changes a time function $x(t)$ into a frequency domain representation $X(f)$. Once in the frequency domain, it is easy to analyze the "frequency content" of a signal. Plotting the Fourier transform allows us to visually determine the relative proportion of different frequencies present in the input signal (the x-axis is now frequency, not time). For example, the transform of a sine wave would look like a single spike, indicating that only one frequency was present. The transform of a circuit's output that's plagued with noise from fluorescent lights would probably have a noticeable peak at 60Hz. Similarly, the transform of the voice of a high-pitched opera singer would likely be concentrated around 8–10KHz.

The continuous-time Fourier transform and the inverse transform formulas are given below:

Fourier Transform	Inverse Fourier Transform
$X(f) = \int_{-\infty}^{\infty} x(t)e^{-j2\pi ft}dt$	$x(t) = \int_{-\infty}^{\infty} X(f)e^{j2\pi ft}df$

Some textbooks define the Fourier transform in terms of ω (rad/sec), instead of f (Hz). We used ω when describing Bode plots because it made things easier to draw; however, it is more intuitive and simpler to use f with Fourier transforms. In any case, you can convert between the two by using the formula $\omega = 2\pi f$.

Look at the Fourier transform formula. Although it seems complex (no pun intended), the transform of many common time functions are easily found by looking them up in a table. However, to better understand what these formulas actually mean, it is useful to look at the inverse transform formula expressed as a Riemann sum:

$$x(t) \approx \{\ldots\ldots + X(f_0)e^{j2\pi f_0 t} + X(f_1)e^{j2\pi f_1 t} + \ldots\ldots + X(f_\infty)e^{j2\pi f_\infty t}\}\,\Delta f \qquad \text{where } \Delta f = f_{i+1} - f_i \text{ for all } i$$

The inverse formula says that any time function $x(t)$ can be represented as the weighted sum (integral) of many different complex exponentials (sinusoids). There is a different weight $X(f)$ for each different frequency sinusoid. These coefficients $X(f)$ are in general complex, giving the sinusoid a magnitude and a phase. Since $X(f)$ is a complex quantity, the Fourier transform must be graphed or displayed as two separate parts – either real/imaginary or magnitude/phase. For a quick sketch of the frequency content of a signal, often just the magnitude plot is shown. While this is often sufficient for practical purposes, keep in mind that this is only half the picture. Both the magnitude and phase information must be stored if the original signal is to be recovered through an inverse transform. When learning about the Fourier transform, it is important to maintain an intuitive understanding of its significance; otherwise, you will likely just get buried in mathematical details and will fail to understand its practical applications.

16.2 Relationship to Bilateral Laplace Transform

A similarity between the Fourier transform and the bilateral Laplace transform can be noticed immediately when the two formulas are placed side by side.

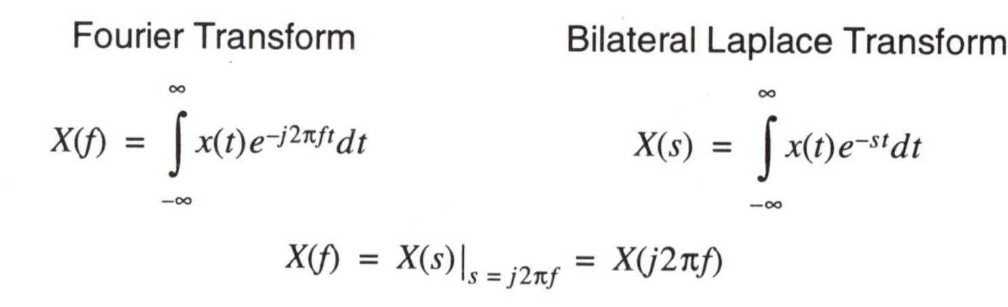

$$X(f) = \int_{-\infty}^{\infty} x(t)e^{-j2\pi ft}dt \qquad X(s) = \int_{-\infty}^{\infty} x(t)e^{-st}dt$$

$$X(f) = X(s)\big|_{s = j2\pi f} = X(j2\pi f)$$

The Fourier transform is equivalent to the bilateral Laplace transform evaluated along the $j\omega$-axis ($s=j2\pi f$). However, this is only valid if the $j\omega$-axis is in the region of convergence (ROC) of the signal's Laplace transform. For example, $e^{2t}u(t)$ has a Laplace transform, but does not have a Fourier transform. Also notice that when referring to Fourier transforms, we drop the "$j2\pi$" and denote $X(j2\pi f)$ as $X(f)$.

One might ask, "Why do we use the Fourier transform when it seems to be just a subset of the bilateral Laplace transform?" It can be shown that any bounded function that has a Laplace transform will have a region of convergence that includes the $j\omega$-axis. In that case, it is often easier to speak about the Fourier transform (no need to deal with ROC). Also, it can be shown that bounded functions that exist for all time, like sinusoids, have a Fourier transform, but do not have a Laplace transform. The Fourier transform extends the class of signals that we can analyze in the frequency domain.

Previously, we said that the Bode plot is a graph of the Laplace transform evaluated along the $j\omega$-axis. Now, we're saying that the Fourier transform is the Laplace transform evaluated along the $j\omega$-axis. What's the difference? Bode plots are generally used to describe *systems*. Fourier transforms are generally used to describe *signals*. But yes, they are virtually the same. Bode plots are magnitude and phase plots on a log scale. Fourier transforms are generally described as $Re\{X(f)\}$ and $Im\{X(f)\}$ plotted on a linear scale. The logarithmic magnitude/phase plot of the Fourier transform of a system's impulse response $h(t)$ is exactly the same as drawing a Bode plot of the system function $H(s)$. Read that last sentence again slowly.

16.3 Fourier Transform Symmetry

Based upon various properties of the signal $x(t)$, there are a few predictable symmetries and characteristics of its associated Fourier transform $X(f)$. For our discussions, we will assume $x(t)$ is real. If you find yourself needing the properties of the Fourier transform of an imaginary signal, just pretend it's real and multiply the transform by j at the end.

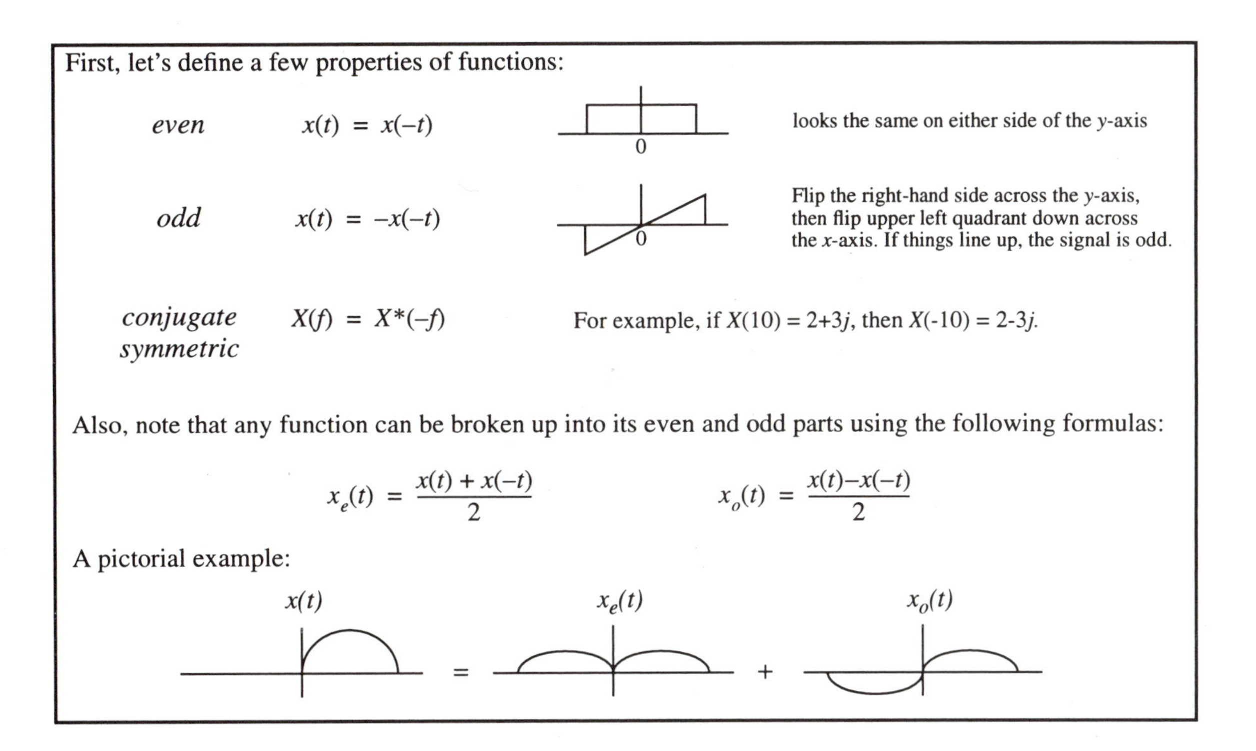

First, let's define a few properties of functions:

even $\quad x(t) = x(-t)$ — looks the same on either side of the y-axis

odd $\quad x(t) = -x(-t)$ — Flip the right-hand side across the y-axis, then flip upper left quadrant down across the x-axis. If things line up, the signal is odd.

conjugate symmetric $\quad X(f) = X^*(-f)$ — For example, if $X(10) = 2+3j$, then $X(-10) = 2-3j$.

Also, note that any function can be broken up into its even and odd parts using the following formulas:

$$x_e(t) = \frac{x(t) + x(-t)}{2} \qquad x_o(t) = \frac{x(t) - x(-t)}{2}$$

A pictorial example:

The table of basic Fourier transform values/symmetries and their proofs are shown below:

x(t) (real)	***X(f) values***	***X(f) symmetry***
anything	complex	conjugate symmetric
even	purely real	even
odd	purely imaginary	odd

PROOFS: (assume $x(t)$ is real)

conjugate symmetry

$$X(f) = \int_{-\infty}^{\infty} x(t)e^{-j2\pi ft}dt = \left[\int_{-\infty}^{\infty} x(t)e^{j2\pi ft}dt\right]^* = X^*(-f)$$

even/odd

$$X(f) = \int_{-\infty}^{\infty} x(t)e^{-j2\pi ft}dt = \int_{-\infty}^{\infty} x(t)\left[\cos(-2\pi ft) + j\sin(-2\pi ft)\right]dt$$

$$X(f) = \int_{-\infty}^{\infty} x(t)\cos(2\pi ft)\,dt - j\int_{-\infty}^{\infty} x(t)\sin(2\pi ft)\,dt$$

If $x(t)$ is even, then the sine term integrates to zero (sine is odd, $x(t)$ is even, even times odd is odd, odd functions integrate to zero on a symmetric interval). This leaves the cosine term which is purely real and depends on f through a cosine, making $X(f)$ purely real and even.

Similarly, if $x(t)$ is odd, the cosine term integrates to zero, leaving an $X(f)$ that is purely imaginary and odd.

In summary, learn the following items about Fourier transforms. The first few are simply restatements of conjugate symmetry, so there really isn't that much to remember.

If $x(t)$ is real...

- $X(f)$ is always conjugate symmetric
- $|X(f)|$ is always even
- $\angle X(f)$ is always odd
- $Re\{X(f)\}$ is always even
- $Im\{X(f)\}$ is always odd
- the even part of $x(t)$ transforms to the real part of $X(f)$
- the odd part of $x(t)$ transforms to the imaginary part of $X(f)$

Also, keep in mind...

- $magnitude(a+b) \neq magnitude(a) + magnitude(b)$ [magnitude is not a linear operation]
- $phase(a+b) \neq phase(a) + phase(b)$ [phase is not a linear operation]

16.4 Fourier Transform Properties

In addition to symmetry there are various other properties associated with the Fourier transform. These properties are easily derived from manipulations of the forward and inverse Fourier transform formulas. See Section 4.6 of *Signal and Systems* by Oppenheim et al (1983) if you want to see the derivations or just need more of an explanation. The following table lists the most common properties of the relationship between the time and frequency domains.

Property	$x(t)$	$X(f)$
Linearity	$ax_1(t) + bx_2(t)$	$aX_1(f) + bX_2(f)$
Duality	$X(t)$	$x(-f)$
Convolution	$x(t) * w(t)$	$X(f)W(f)$
Product	$x(t)w(t)$	$X(f) * W(f)$
Time Shift	$x(t-t_o)$	$e^{-j2\pi f t_o}X(f)$
Frequency Shift	$e^{j2\pi f_o t}x(t)$	$X(f-f_o)$
Differentiation	$\frac{dx(t)}{dt}$	$j2\pi fX(f)$

Property	$x(t)$	$X(f)$
Times t	$tx(t)$	$-\frac{1}{j2\pi}\frac{dX(f)}{df}$
Time Scaling	$x(at)$	$\frac{1}{\lvert a\rvert}X(f/a)$

The duality property is a bit subtle and deserves further discussion. Start with a signal $x(t)$ and take its transform to produce $X(f)$. Then, pretend that signal is a time signal $X(t)$ and take its transform again. You will produce something that is identical to the signal you started with, except it is time-reversed. This means that if you build a circuit to take Fourier transforms, you can use the same circuit for taking inverse transforms, as long as you are careful about time-reversal.

Although it is probably pretty clear from the table, let's also reiterate the *convolution* and *product* properties. The transform of the convolution of two signals is the product of the individual transforms. Similarly, the transform of the product of two signals is the convolution of the individual transforms. Just to be sure you got it:

$y(t) = h(t) * x(t)$	$\Leftrightarrow$	$Y(f) = H(f)X(f)$
convolution in the time domain	$\Leftrightarrow$	multiplication in the frequency domain
$y(t) = h(t)x(t)$	$\Leftrightarrow$	$Y(f) = H(f) * X(f)$
multiplication in the time domain	$\Leftrightarrow$	convolution in the frequency domain

16.5 Parseval's Theorem and More

Here are a few more properties of Fourier transforms that you will find useful:

Area in Time (DC offset)

$$X(0) = \int_{-\infty}^{\infty} x(t)dt$$

This is nothing more than the Fourier transform formula evaluated when $f=0$. Think of $X(0)$ as the area under the time function $x(t)$.

Area in Frequency

$$x(0) = \int_{-\infty}^{\infty} X(f)df$$

As expected, the same sort of relation holds for the inverse Fourier transform formula. Think of $x(0)$ as the area under the function $X(f)$.

Parseval's Theorem

$$\int_{-\infty}^{\infty} \lvert x(t)\rvert^2 dt = \int_{-\infty}^{\infty} \lvert X(f)\rvert^2 df$$

Parseval's theorem equates the area under the magnitude squared of $x(t)$ and $X(f)$. In other words, the "energy" in the time domain equals the "energy" in the frequency domain.

16.6 Basic Fourier Transform Pairs

Here is a table illustrating some of the more common Fourier transform pairs:

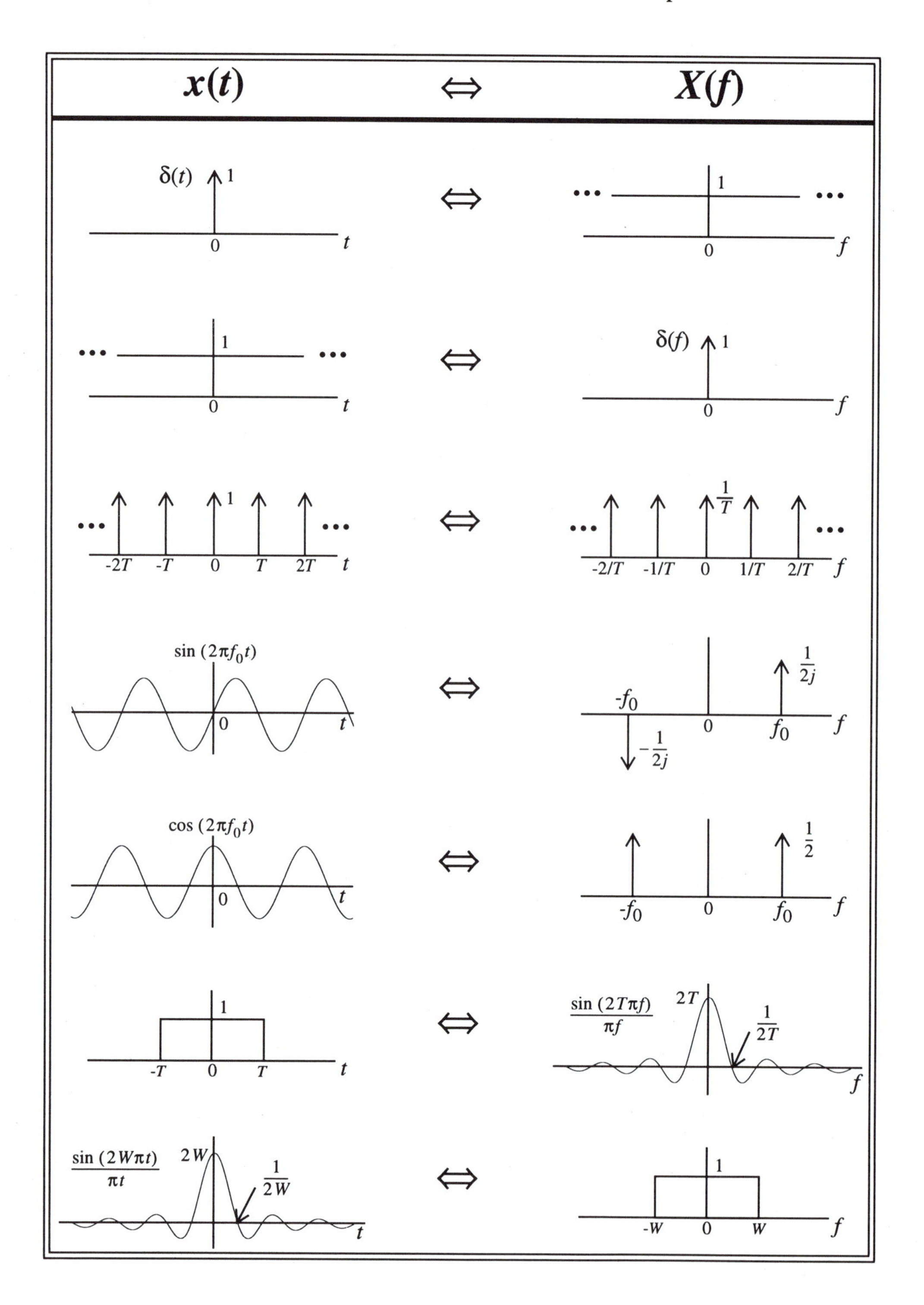

In this book, we will refer to a sinc function as anything of the form $\sin(x)/x$. The box↔sinc Fourier transform pairs always seem to give people the most difficulty. They are easy to remember if you notice the following facts. Practice a few box↔sinc transforms with your friends until you feel comfortable with them.

When finding the sinc:

- The height of the sinc is always equal to the area of the box (see Section 16.5).
- The first zero crossing of the sinc wave is at 1/(total width of box).
- The sinc function is always of the form $\frac{(\text{height of box})\sin(\text{width of box} \cdot \pi X)}{\pi X}$ where X is either t or f.

When finding the box:

- If given the equation of the sinc function, just read off the height and width of the box using the template equation given above.
- If given the graph of a sinc function, the width of the box is equal to 1/(first zero crossing point of the sinc) and the box height is equal to the area of the sinc function (see Section 16.5). The total area underneath a sinc function is equal to the area of the triangle formed by taking the following three points: the peak of the sinc, the first negative zero crossing, and the first positive zero crossing.

16.7 Duration-Bandwidth and the Uncertainty Principle

All Fourier transform pairs are constrained by the *uncertainty principle.* This concept states that it is impossible to define a signal that is arbitrarily small in both the time and frequency domains. In other words, a signal of short duration in the time domain must have a wide Fourier transform. In fact, after appropriately defining the *duration* (time domain) and the *bandwidth* (frequency domain) of a signal in terms of its statistical moments, it is possible to prove that all real waveforms must satisfy the following compact relationship.

$$(\text{duration})(\text{bandwidth}) \geq \frac{1}{\pi}$$

This statement of the uncertainty principle is largely of theoretical significance; see Chapter 16 of *Circuits, Signals, and Systems* by Siebert (1986) for its derivation. What is really important however is to remember the following implications of this concept.

- Signals "narrow" in the time domain (e.g. impulse) are "wide" in the frequency domain (flat line).
- Signals "wide" in the time domain (e.g. flat line) are "narrow" in the frequency domain (impulse).
- Signals finite in the time domain (e.g. a box) are infinitely long in the frequency domain (sinc).
- Signals infinitely long in the time domain (e.g. sinc) are finite/bandlimited in the freq. domain (box).

It is interesting to note that Gaussian functions of the form e^{-t^2} have the smallest duration-bandwidth product. The uncertainty principle allows us to answer basic questions like, "Which signal has a higher bandwidth: $\sin(t)/t$ or $\sin(2t)/(2t)$?" Answer: the second one.

16.8 Fourier Transforms of Discrete Signals

It is also possible to take the Fourier transform of a discrete signal. The result is known as the discrete-time Fourier transform (DTFT). The DTFT, like its continuous-time counterpart, is a continuous function of frequency. However it is more common to discuss the DFT, which is just equally spaced samples of the DTFT and can be implemented on a computer. A computationally efficient algorithm has been designed to speed the processing of DFT's; this algorithm is known as the much celebrated Fast Fourier Transform, or FFT for short. It is probably one of the most commonly used algorithms in the world today. See Chapter 5 of *Discrete-Time Signal Processing* by Oppenheim and Schafer (1989) for a complete discussion of Fourier transforms for discrete-time signals.

16.9 A Sample Problem

Given the following real signal, find:

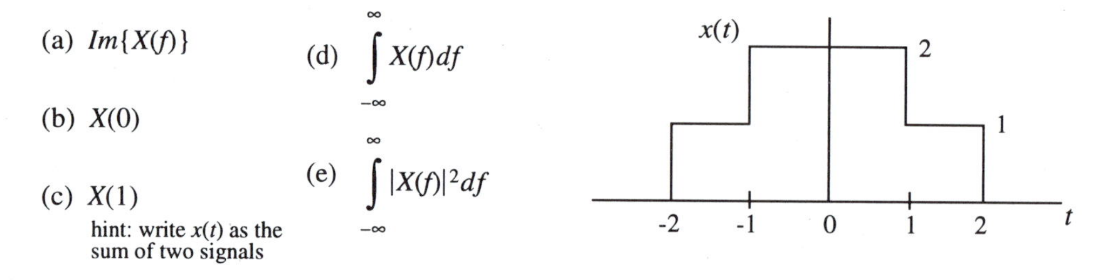

(a) $Im\{X(f)\}$

(b) $X(0)$

(c) $X(1)$
hint: write $x(t)$ as the sum of two signals

(d) $\int_{-\infty}^{\infty} X(f)df$

(e) $\int_{-\infty}^{\infty} |X(f)|^2 df$

Answers:

(a) 0 (b) 6 (c) 0 (d) 2 (e) 10

CHAPTER 17 Filters

Overview

Filtering is the process of altering the frequency content of a signal. It is probably the most widely used signal processing operation. Filters are ubiquitous in the world around you, with the most common examples being in your home stereo system. This chapter describes the various classes of filters and the processes involved in designing them. Practical considerations such as phase distortion, digital filters, and switched-capacitor implementations are also discussed. After reading this chapter the phrase, "Crank the bass, man!" should take on a whole new meaning.

17.1 What is Filtering?

Recall that the Fourier transform shows the frequency content of a signal. Filtering is the process of selectively removing or altering parts of this frequency content to create a new signal. A common type of filter that you should all be familiar with are the bass and treble knobs on some basic stereo systems. When turning the knobs, you are altering the frequency content of the audio signal by boosting or reducing high or low frequencies.

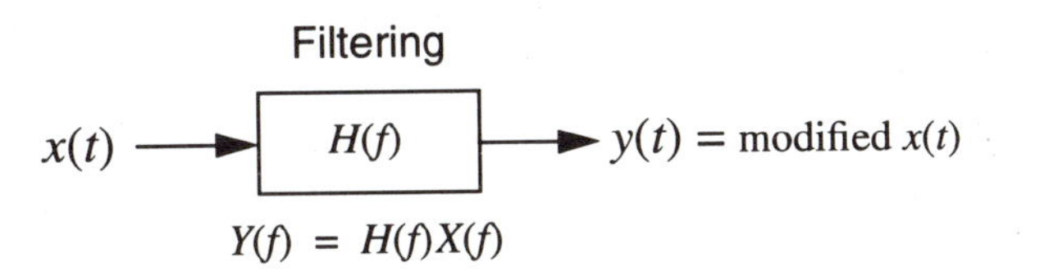

$Y(f) = H(f)X(f)$

Choose shape of $H(f)$ to appropriately alter $X(f)$ in desired manner.

The box containing $H(f)$ is known as the filter.

17.2 Types of Filters

Filters, both in continuous and discrete-time, can be grouped loosely into one of four categories: lowpass, highpass, bandpass, and notch, which are illustrated in the following table. Note that only the positive side of the frequency axis is drawn; the complete Fourier transform magnitude is symmetric since the filter is real. We will initially only discuss the magnitude of the filter; phase response will be addressed in Section 17.9. The rationale behind the shapes of the filters is better understood if you keep in mind that the output of the filter is the input $X(f)$ *multiplied* by $H(f)$. The idealized filters multiply the to-be-removed-sections by zero. The names of the different types of filters are also quite intuitive – a lowpass filter passes low frequencies, but stops high frequencies. A bandpass filter passes a band of frequencies, but stops other frequencies.

Filter Types

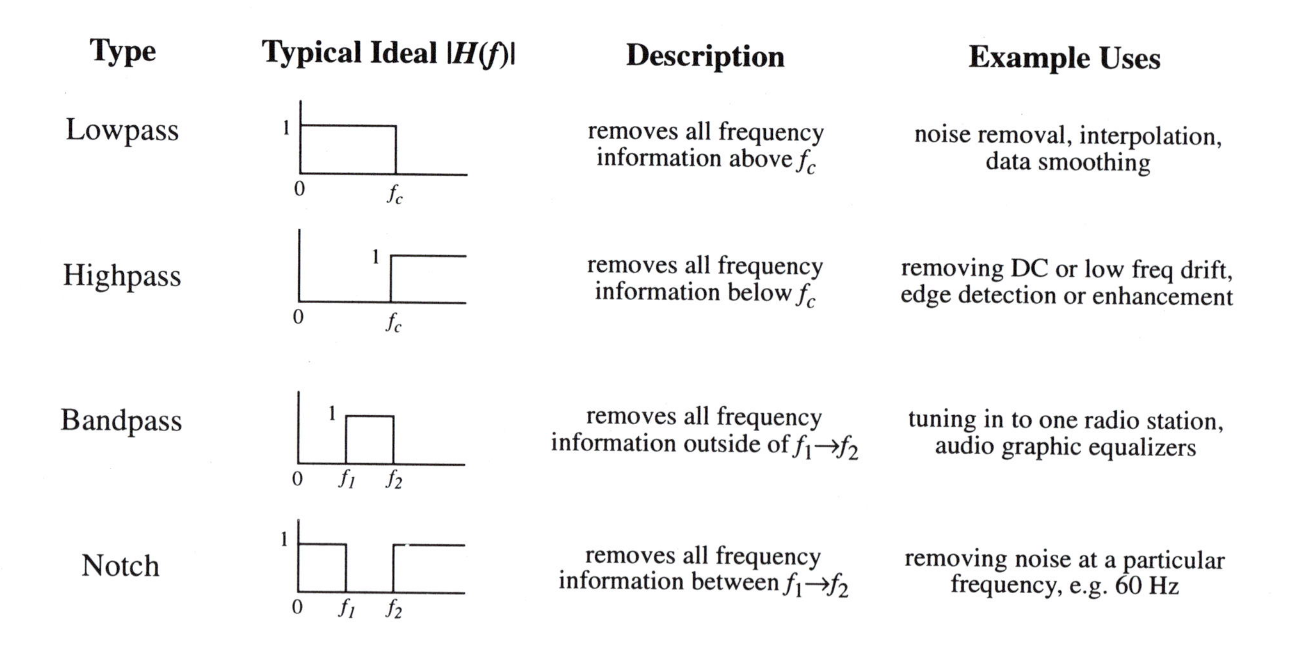

Type	Typical Ideal $\|H(f)\|$	Description	Example Uses
Lowpass	1, 0, f_c	removes all frequency information above f_c	noise removal, interpolation, data smoothing
Highpass	1, 0, f_c	removes all frequency information below f_c	removing DC or low freq drift, edge detection or enhancement
Bandpass	1, 0, f_1, f_2	removes all frequency information outside of $f_1 \rightarrow f_2$	tuning in to one radio station, audio graphic equalizers
Notch	1, 0, f_1, f_2	removes all frequency information between $f_1 \rightarrow f_2$	removing noise at a particular frequency, e.g. 60 Hz

17.3 Non-Ideal Filters (the real world)

The idealized filters shown in Section 17.2 are exactly that – ideal. However, they cannot exist in the real world. To help understand why, let's take the ideal lowpass filter as an example. The magnitude of its frequency response is shown below:

Ideal Lowpass Filter

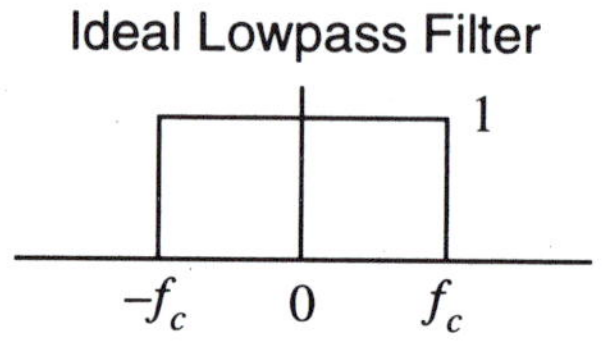

Recall from Section 16.7 in the Fourier transform chapter that any function that is of finite length in the frequency domain must be of *infinite* length in the time domain. Because $H(f)$ is finite, this means that the corresponding $h(t)$ could not possibly be causal ($h(t)=0$ for $t<0$) or even made to be causal by shifting it in time, since it is of infinite length. Non-causal systems cannot be realized by physical systems (e.g. a circuit). If we try to avoid the real world and implement this filter digitally, where causality is not much of a concern (since you can easily talk about negative time), completely ideal filters are still not possible because of their infinite length. The system would require an infinitely long discrete time convolution to perform the filtering. See Chapter 15 of *Circuits, Signals, and Systems* by Siebert (1986) for more information on the issues concerning non-ideal filters.

17.4 Filter Terminology

Since we know that ideal filters cannot exist, then what do the frequency responses of real filters actually look like? Basically, unlike their idealized versions, they are not flat, nor are their transitions perfectly sharp. The general form of an non-ideal lowpass filter is shown below:

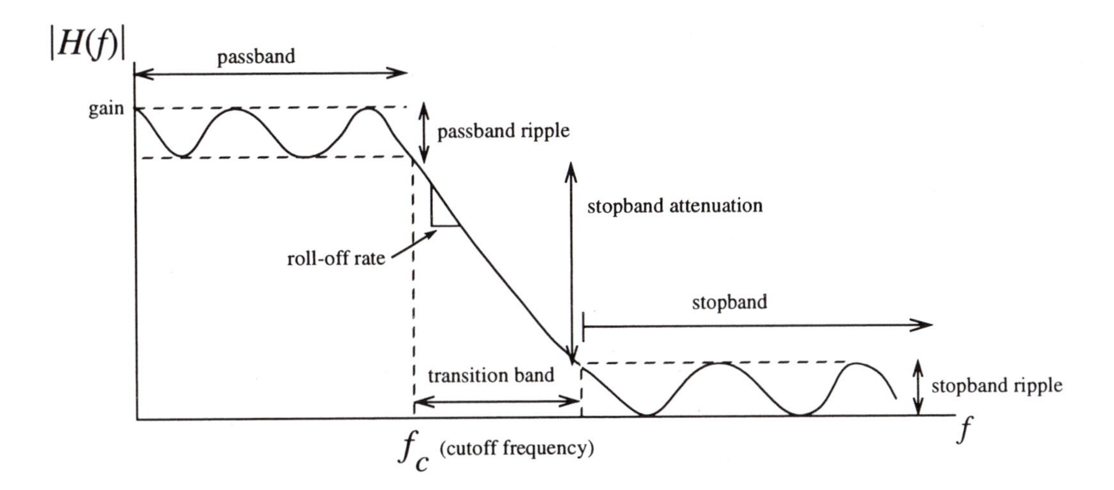

- ***Passband*:** the frequency range you are interested in *preserving* in the output signal
- ***Stopband*:** the frequency range you are interesting in *eliminating* in the output signal
- ***Transition-band*:** the band over which the frequency response transitions from the passband to the stopband; would like it to be as small as possible
- ***Gain*:** refers to the amount of maximum amplification of the signal in the passband
- ***Stopband attenuation*:** the difference in dB between the passband gain and stopband gain
- ***Passband ripple*:** the maximum fluctuation in filter's frequency response in the passband; usually measure in dB
- ***Stopband ripple*:** the maximum fluctuation in the frequency response in the stopband; essentially irrelevant as long as the stopband attenuation is met
- ***Roll-off rate*:** the steepness of the slope in the transition band; usually multiples of 20dB/decade (note 20dB/decade = 6dB/octave, where an octave is a doubling in frequency)
- ***Order*:** the number of poles in the system function $H(s)$; the higher the order, the steeper the roll-off rate and the shorter the transition band; higher order filters are more complicated to build
- ***Cutoff frequency*:** the edge of the passband; also known as the corner frequency or 3dB point since it is the "corner" in the asymptotic Bode plot and is generally 3dB lower than the peak passband gain
- ***Q*:** the sharpness of the peak in a bandpass filter; defined as center frequency divided by the half-power bandwidth (from 3dB point to 3dB point); a measure of how close the poles are to $j\omega$-axis

17.5 Designing a Continuous-Time Filter

Designing a filter involves nothing more than finding a system function $H(s)$ that has the desired frequency response (Bode plot). This translates to placing the poles and zeros of $H(s)$ in the appropriate place in the s-plane in order to achieve the desired shape of $H(f)$. How many poles? How many zeros? Where should they go? Don't panic! Someone has already come up with a set of polynomials (depending on what order filter you want) whose roots are the appropriate pole locations. For reasons that will become apparent shortly, we will initially focus only on lowpass filter design. The design procedure is to first pick your cutoff frequency, the desired filter order, and the "type" of lowpass filter (see below). A computer is usually used to determine the roots (pole/zero locations) of the appropriately chosen polynomials, which can be found in virtually all filter design or op-amp textbooks (e.g. *Operational Amplifier Circuits: Theory and Applications* by Kennedy (1988)). Depending on your computer program, you may have to manually pick out only the roots in the left half plane to insure stability. Voila, you have yourself a filter. There are many different "types" of lowpass filters depending on your requirements. Three of the most common are Butterworth, Chebyshev, and Elliptical and are described below.

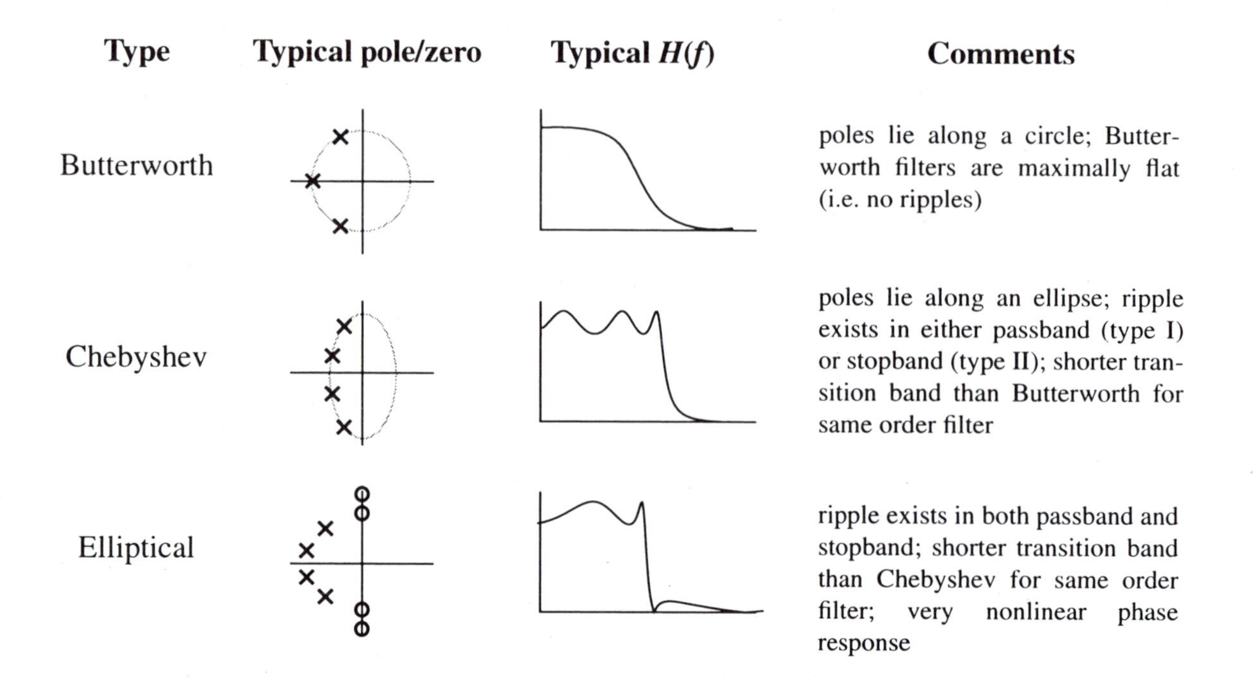

Designing Other Types of Filters

What about designing filters that are not lowpass? No problem. First, design a lowpass $H(s)$ that has a cutoff frequency of $\omega_c = 1$ and then apply the mathematical transformations shown in the following table, replacing "s" with the appropriate expression. This procedure can be used to generate highpass, bandpass, or notch filters with arbitrary cutoff or center frequencies.

If $H(s)$ is a lowpass filter with cutoff frequency $\omega_c = 1$, then...

$$\text{Highpass} = H(\omega_c/s)$$

ω_c = desired cutoff of highpass filter

$$\text{Bandpass} = H\left(\frac{s^2+\omega_0^2}{s \cdot BW}\right)$$

$$\text{Notch} = H\left(\frac{s \cdot BW}{s^2+\omega_0^2}\right)$$

ω_0 = desired center frequency of bandpass or notch filter

$BW = \omega_h - \omega_l$ = desired width of bandpass or notch filter

17.6 Circuit Realizations of Filters

Now that we have the desired pole locations, how do we actually build a filter? Recall that impedances of basic circuit elements can be pieced together to create arbitrary transfer functions. The goal now is to come up with the correct placement and values of resistors, capacitors, inductors, and possibly op-amps to construct the desired filter's $H(s)$. There are two broad classifications of filter types: active and passive.

Passive filters consist of purely unpowered circuit elements, such as R's, L's, and C's. Some advantages of passive filters are that they don't need a power supply, and they can be quite cheap to build if the filter is only first or second order and high accuracy is not needed. Disadvantages include: the difficulty of finding the exact valued capacitor or inductor for accurate pole placement; being unpowered means the filter cannot provide any sort of signal gain; and their input/output impedances are not infinite/zero, thus sometimes causing loading problems when used as part of a larger circuit. For example, the circuits shown below illustrate the effect of cascading two first-order passive RC filters. The circuit on the left has a cutoff frequency of 10 rad/sec. By stringing two of them together, can we make a second-order lowpass filter with the same cutoff frequency? No! It's probably a good exercise to find the system function of the second circuit if you're not convinced. The problem is that the input impedance of the second RC stage is nonzero, thus stealing current from the first stage and causing it to behave differently than if it were unloaded.

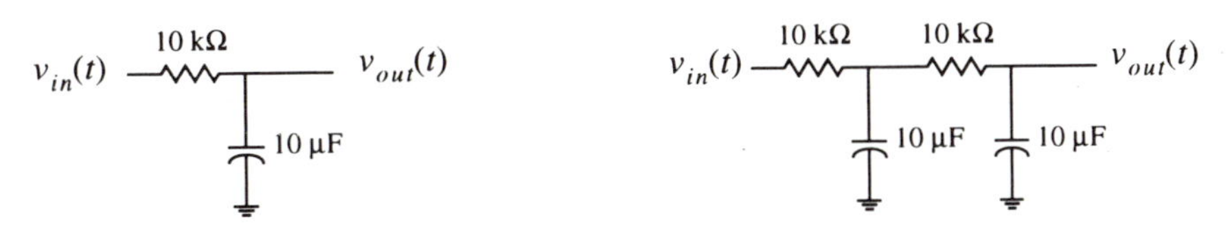

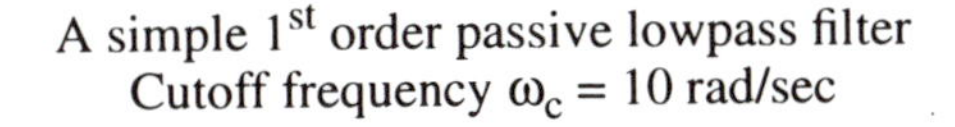
A simple 1st order passive lowpass filter
Cutoff frequency ω_c = 10 rad/sec

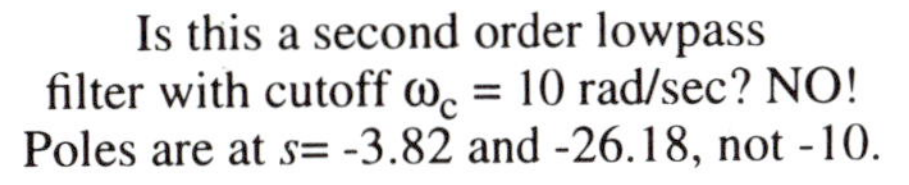
Is this a second order lowpass filter with cutoff ω_c = 10 rad/sec? NO!
Poles are at s= -3.82 and -26.18, not -10.

Active filters on the other hand require an external power supply in order to maintain or to boost signal strength. Advantages include: the ability to amplify signals; the absence of often expensive and bulky inductors; and an essentially infinite input impedance (or near zero output impedance). Some disadvantages include the possible introduction of extra noise in the signal and the fact that the maximum operating frequency of the filter is constrained by the gain-bandwidth product of the amplifying element.

There are several common active filter circuits that have already been designed; one merely needs to plug in the appropriate valued components. One of the more common templates is known as the Sallen-Key lowpass filter, which can implement an arbitrary second-order transfer function. Higher order filters can then be obtained by cascading a series of Sallen-Key circuits; loading effects are eliminated since the ideal op-amp has zero output impedance. Exchanging the position of the R's and C's in the lowpass Sallen-Key shown below will produce a second-order system with two zeros at the origin, which is a highpass filter. There are also forms for a Sallen-Key bandpass implementation. For more information, refer to Chapter 5 of *The Art of Electronics* by Horowitz and Hill (1989).

Sallen-Key Lowpass Filter Circuit

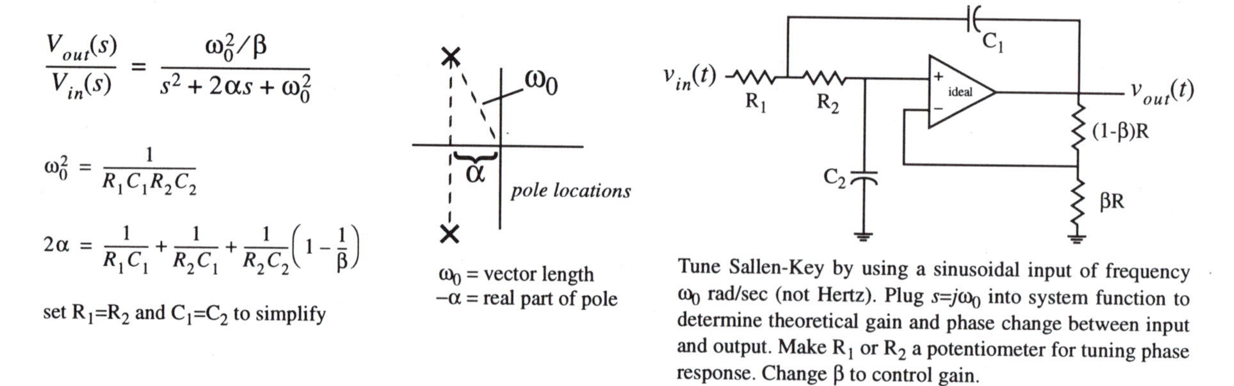

$$\frac{V_{out}(s)}{V_{in}(s)} = \frac{\omega_0^2/\beta}{s^2 + 2\alpha s + \omega_0^2}$$

$$\omega_0^2 = \frac{1}{R_1 C_1 R_2 C_2}$$

$$2\alpha = \frac{1}{R_1 C_1} + \frac{1}{R_2 C_1} + \frac{1}{R_2 C_2}\left(1 - \frac{1}{\beta}\right)$$

set $R_1=R_2$ and $C_1=C_2$ to simplify

ω_0 = vector length
$-\alpha$ = real part of pole

Tune Sallen-Key by using a sinusoidal input of frequency ω_0 rad/sec (not Hertz). Plug $s=j\omega_0$ into system function to determine theoretical gain and phase change between input and output. Make R_1 or R_2 a potentiometer for tuning phase response. Change β to control gain.

17.7 Recognizing a Filter

Given a pole/zero diagram or the transfer function equation of a filter, it is relatively straightforward to guess its type (a common multiple choice test question). The way to solve these problems is to make a mental sketch of the Bode plot by tracing your finger up the $j\omega$-axis starting at the origin and thinking about a large rubber sheet lying over the pole/zero plot (see Section 6.7). Or, if you're given just the equation of the filter transfer function, one thing that will help is to try plugging in $s=0$ and $s=\infty$ and evaluating $|H(s)|$. Just for practice, classify each of the following s-plane pole/zero plots as coming from lowpass, highpass, bandpass, or notch filters.

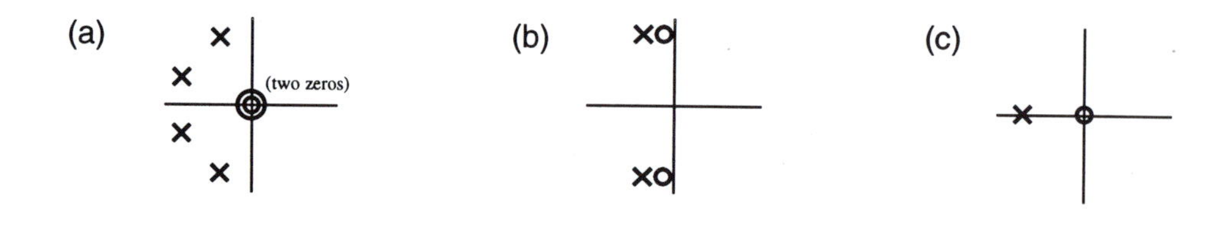

(a) bandpass - it looks like a 4th order lowpass with two zeros added, which kill the DC
(b) notch filter - zeros on the $j\omega$-axis remove a particular frequency, poles support others
(c) highpass - draw the Bode plot

17.8 A Visual Example

The following plots are a good visual demonstration of the effects of various types of filters. The input signal is a sinusoid whose frequency is linearly increasing over time. This type of signal is often referred to as a "chirp" signal since if you listened to it repeatedly, it would sound vaguely like a bird's chirping sound. The four other plots show the output of four different types of filters when presented with this input chirp signal. The non-zero width of the transition band is clearly evident here.

Input "Chirp" Signal x(t)

Lowpass Filter Output y(t)

Bandpass Filter Output y(t)

Highpass Filter Output y(t)

Notch Filter Output y(t)

17.9 Phase Response

Up until now, we have been emphasizing the magnitude response of filters. But this is only half the picture. Recall that a frequency response consists of two parts –magnitude and phase. We know what an ideal magnitude response looks like, but what is an ideal phase response? Ideally, a filter should have a *linear* phase response, meaning that the plot of $\angle H(f)$ vs. f should be a straight line. Why? Imagine a sinusoidal input to this filter; the output sinusoid will be of the same frequency, but possibly with a different magnitude and phase. This phase difference can be translated into a time delay as shown below. Remember, delayed output signals appear as a negative phase difference.

$$\text{input} = A\sin(2\pi ft + \phi_1)$$
$$\text{output} = B\sin(2\pi ft + \phi_2)$$

$$\underset{\text{(negative = delay)}}{\text{time diff (sec)}} = \frac{\text{phase difference (radians)}}{\text{sinusoid frequency (rad/sec)}} = \frac{\phi_2 - \phi_1}{2\pi f}$$

Now, if the phase difference is a linear function of the sinusoid frequency, then the frequency cancels in the numerator and denominator and the time difference is merely a constant. This means that the output is merely a constant time delay difference from the input, for all possible input frequencies. If the phase response was not linear, then different frequencies would be delayed by different amounts. For example, after a filter the drums might be heard before the opera singer's voice, instead of at the same time. The signal would sound "separated" or "distorted." Find the specs of your favorite hi-fi stereo system. You should see something called "total linear phase distortion: 0.1%." This figure is a measure of the departure from linearity for the combined phase responses of all the internal filters. Also remember, the phase response need only be linear in the passband.

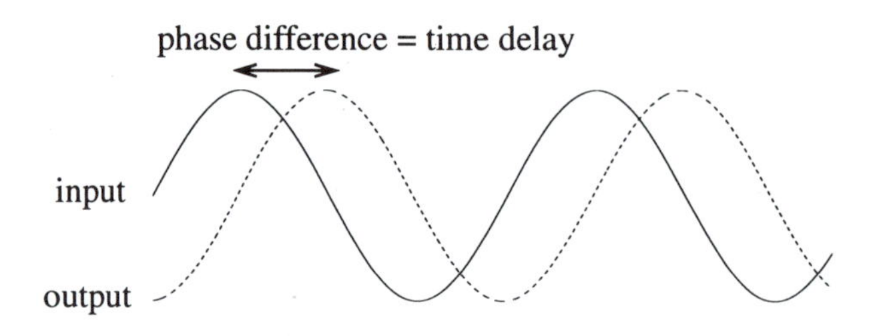

Butterworth, Chebyshev, and Elliptic have good amplitude responses, but have relatively poor phase responses (elliptical is the worst of the three and Chebyshev type II is the best). For most applications where the signal of interest is mostly within a narrow frequency band, you probably won't notice the phase distortion. If the signal is audio, you will also find that the human ear is quite insensitive to phase distortion when it comes to intelligibility. However, if the phase response is critical, like in high quality recordings, you would be better off using other types of filters such as the parabolic or the Thompson (bessel). These filters have a more linear phase response and a better time-domain step response (no ringing), but at the expense of a more gradually sloping transition band.

Another class of filters we have neglected to discuss is known as the *allpass* filter. Ideal allpass filters have a magnitude response that is flat over all frequencies. So what good is it? It's used to modify the phase of an input signal. A typical use is correcting the linear phase distortion encountered with other types of filters.

17.10 Digital Filters

Digital filters operate on discrete-time signals. They have a transfer function $\tilde{H}(z)$, cutoff frequencies, and all of the other characteristics that define a continuous-time filter. While analog filters operate by internally scaling and integrating their signals, digital filters work by scaling and summing (recall the difference between continuous-time convolution and discrete-time convolution). Some advantages of digital filters are: their characteristics are truly time-invariant (not susceptible to temperature changes, aging components, etc.); filter behavior can be easily changed just by reprogramming your computer instead of having to rewire a circuit; and the filtering process is essentially noise-free.

There are basically two types of digital filters: IIR (infinite impulse response) and FIR (finite impulse response), referring to the length of the filter's $h[n]$. As seen in previous sections, continuous-time filter design is a well-established art, full of cookbook formulas and standard filter templates. Therefore, one way to design a digital IIR filter is to design the continuous-time equivalent and simply apply a continuous-time to discrete-time conversion process, such as impulse invariance, forward/backward differences, or the bilinear transformation (see Section 8.5). IIR filters are often referred to as recursive filters since they are described by difference equations where the output $y[n]$ depends not only on previous values of the input $x[n]$, but also previous values of the output.

FIR filters, as the name indicates, are finite in length. The output value $y[n]$ depends only on a fixed number of previous values of the input $x[n]$. The advantage of using an FIR filter is that this type of filter always has a linear phase response. One way of designing an FIR filter is to design the desired digital IIR filter and simply truncate the impulse response by multiplying it by a window. For arbitrary filter shapes, the Parks-McClellan algorithm is an iterative optimal equiripple FIR filter design method. Just input the desired shape of the transfer function and the filter order, and out pops the coefficients of the best possible FIR filter. So which should you use? IIR or FIR? The pros and cons of each as well as a thorough discussion on discrete-time filter design can be found in Chapter 7 of *Discrete-Time Signal Processing* by Oppenheim and Schafer (1989).

Digital filters are becoming more and more common everywhere around us, but that doesn't mean analog filters are useless. Given an analog signal such as audio that needs filtering, think about all of the extra overhead required to perform that filtering in the digital domain: an A/D converter, a fast computer or dedicated DSP chip, memory to hold the filter coefficients, source code to actually perform the convolution, and a D/A converter to output the modified signal. Contrast that several hundred dollar setup to a single op-amp filter that can function at several hundred megahertz at a mere fraction of the cost. Of course, if your signal is digital to begin with (the closing Dow Jones Industrial Average stock index, the output of a CCD digital camera, etc.), then it makes sense to use a digital filter. Engineering is all about tradeoffs, and filter design is no exception.

17.11 Switched-Capacitor Filters

A switched-capacitor filter is not the name of a new type of filter, but rather a new methodology of implementing one. It combines the speed and convenience of an analog filter with the flexibility and accuracy of the digital world. The pole locations can be precisely controlled by merely altering the frequency of a digital square wave input signal. The filter works on the notion that a capacitor rapidly switched between two points and ground behaves like a resistor connected between those two same points. Huh? A capacitor can become a resistor? An explanation is in order. Consider the following diagram:

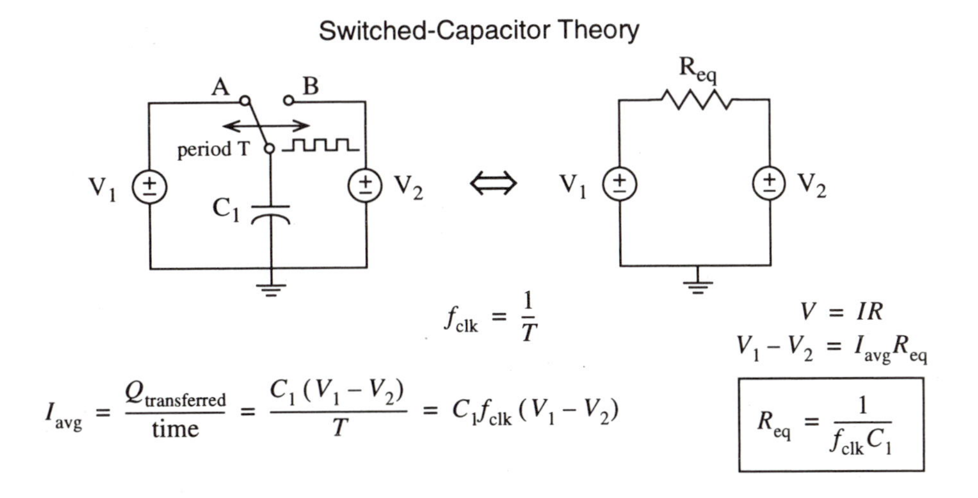

When the switch is in position A, the amount of charge on the capacitor is $Q_A = C_1 V_1$. When the switch is thrown to position B, the amount of charge is supposed to be $Q_B = C_1 V_2$. The difference between these two charge levels is the amount of charge that flows into (or out of) source V_2. In other words, every T seconds (the period of the clock) a charge of $C_1(V_1 - V_2)$ Coulombs is effectively transported from point A to point B. Net charge flowing per unit time is the definition of current. Since we know the voltage difference as well as the average current flowing, we can use Ohm's law to find the equivalent resistance between points A and B. Thus the switching capacitor behaves like a resistor. Big deal! Why don't we just use a resistor and save ourselves the hassle? Things should become clear after the following example:

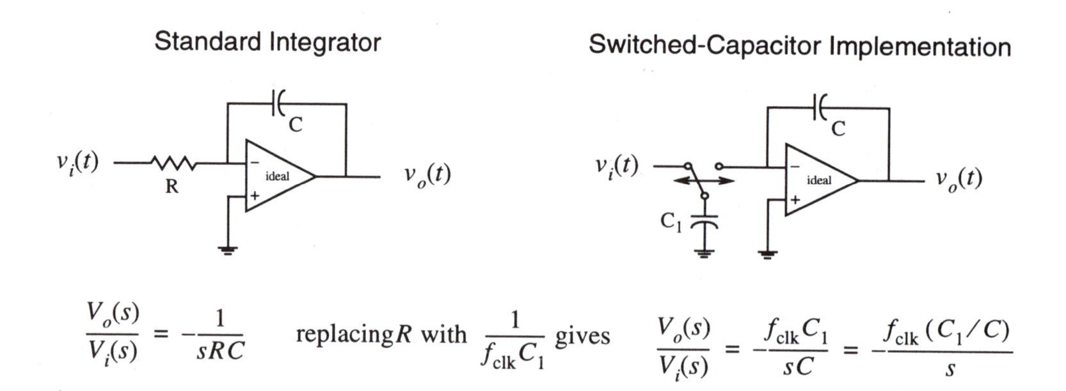

$$\frac{V_o(s)}{V_i(s)} = -\frac{1}{sRC} \quad \text{replacing} R \text{ with } \frac{1}{f_{\text{clk}} C_1} \text{ gives } \quad \frac{V_o(s)}{V_i(s)} = -\frac{f_{\text{clk}} C_1}{sC} = -\frac{f_{\text{clk}} (C_1 / C)}{s}$$

Here we have taken a simple integrator op-amp circuit and shown it in its switched-capacitor implementation. The advantage of the latter is that the transfer function is now dependent only on the *ratio* of two capacitors and is completely independent of their absolute value. Small-valued capacitors are easily implemented directly in silicon, which means the entire filter can be captured directly inside a single chip. Furthermore, the pole placement is guaranteed to be extremely accurate since (1) the input clock frequency can be governed by a readily available precise crystal oscillator, and (2) although a single capacitance value cannot be very well controlled, the ratio of two capacitors can be made very precise since capacitance is dependent on the size of the doped area in silicon (accurate to the sub-micron level during wafer fabrication). The switching mechanism itself is easily implemented by hooking the input clock signal to the gates of MOS transistors to appropriately connect/disconnect the capacitor.

The biggest problem when designing an analog filter is finding precise resistive and capacitive components that don't change in value with temperature, etc. in order to achieve accurate pole placement. The switched-capacitor filter eliminates that problem by making the cutoff frequency dependent solely on the input frequency of a digital square wave. Furthermore, the same chip can perform filtering at a variety of different center or cutoff frequencies by merely altering the frequency of the clock signal.

Nevertheless, you never get something for nothing. Because the current (charge) is transported across the switch in discrete bursts, the output of a switched-capacitor filter tends to look jagged, like a staircase. This effect, however, is minimized when the frequency of the clock signal is *much* greater than the maximum frequency of the input signal. In any case, it is easy to remove the jagged appearance by following the switched-capacitor filter with a simple lowpass filter to smooth out any sharp edges.

CHAPTER 18 Modulation

Overview

Modulation is a part of all modern day electronic communications such as radio, television, and telephony. Modulation is basically the process of moving or "frequency shifting" the spectrum of a signal to a new frequency range, thus allowing multiple signals to coexist on a single transmission medium. This chapter discusses several basic modulation and demodulation schemes.

18.1 What is Modulation?

The easiest way to think about modulation is to use radio stations as an example. The sound that radio stations want to transmit is in the frequency range from 20Hz to 20KHz, the range of human hearing. However, if there is more than one radio station in the area, the signals from the two stations will overlap and will be impossible to distinguish unless they are separated in the frequency domain in some manner. In that case, appropriately placed filters could pick out the desired radio station. Modulation is the process of moving a frequency spectrum to a new center frequency, like 88.1MHz in the case of MIT's college radio station. Demodulation is the process of moving that frequency spectrum back to its original location. Two broad classes of modulation schemes are known as amplitude modulation and frequency modulation.

18.2 Amplitude Modulation

Amplitude modulation, commonly known as AM, is the most basic form of modulation. Here, the input signal $x(t)$, which could be music, video, or any other bandlimited waveform, is multiplied by a sinusoidal carrier signal to produce the modulated output signal $x_{AM}(t)$.

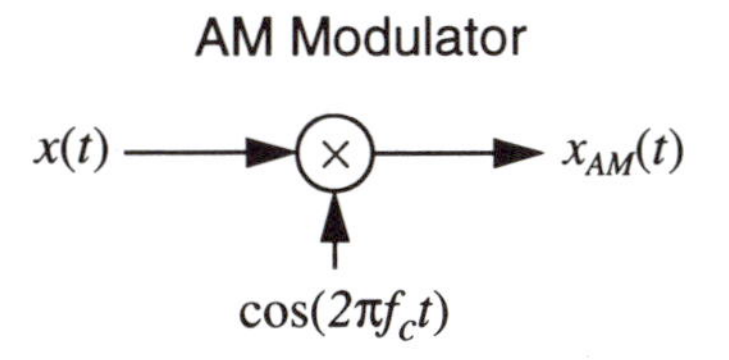

A bandlimited signal is a function whose Fourier transform is zero outside a given range of frequency, i.e. $X(f)=0$ for $|f| > W$. For simplicity, when drawing spectra we will draw only the real part of $X(f)$; however, remember that in general it has both real and imaginary parts. The carrier signal is generally a sinusoid with frequency $f_c \gg W$. The reason for such a high carrier frequency and the mechanics of how modulation achieves frequency shifting will soon be apparent.

Below is a typical "bandlimited" signal $x(t)$ and its accompanying Fourier transform $X(f)$.

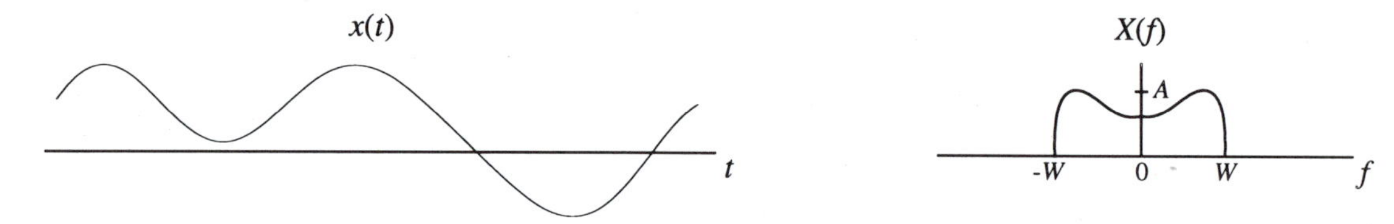

In AM modulation, the signal $x(t)$ is *multiplied* by a cosine in time (the carrier signal). In the frequency domain that translates to *convolving* $X(f)$ with the Fourier transform of a cosine wave. Recall that the Fourier transform of $\cos(2\pi f_c t)$ looks like two impulses in frequency.

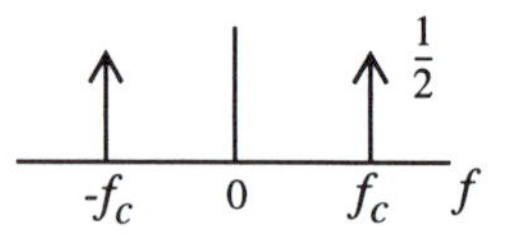

Multiplication in time causes the amplitude of the cosine carrier signal to follow the amplitude of $x(t)$. This is illustrated in the plot of the signal $x_{AM}(t)$ shown below. The effect of the convolution in the frequency domain is to shift one copy of $X(f)$ up to f_c and one copy down to $-f_c$. The spectrum of the modulated signal $X_{AM}(f)$ is drawn below. Note the amplitude is scaled by 1/2 from the cosine spectrum.

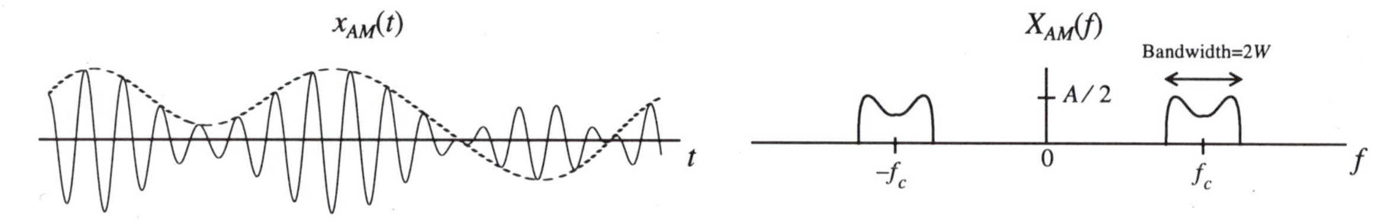

Frequency Multiplexing

It should now seem clear that modulation allows us to combine multiple signals onto a signal transmission channel by simply placing them in different frequency bands. This process is known as frequency multiplexing. For instance, we could combine two signals by AM modulating them with two different carrier frequencies f_1 and f_2 and then summing the results. The Fourier transform of the transmission might look something like the following.

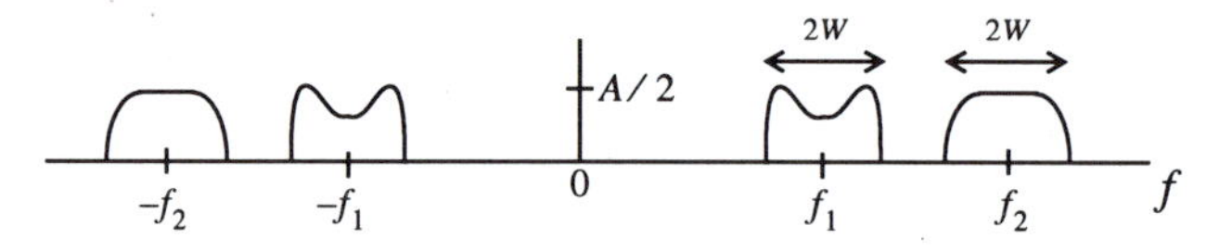

The carrier frequencies f_1 and f_2 must be chosen so that the two pictures do not overlap. As long as this is true, either signal can easily be recovered as we shall soon see.

Synchronous Detector for Demodulation

A synchronous detector is a scheme for demodulating (recovering) an AM signal. The process involves multiplying $x_{AM}(t)$ by the same carrier signal used in the modulator. Note that there must be no phase difference between the modulating and demodulating sinusoids – hence the name synchronous. This time-domain multiplication is a convolution in the frequency domain, which then shifts copies of the spectrum back to zero frequency, among other places. The lowpass filter then isolates and scales the desired copy so that the original spectrum $X(f)$ is fully recovered. The following diagram outlines the overall process.

Synchronous AM Demodulation Scheme

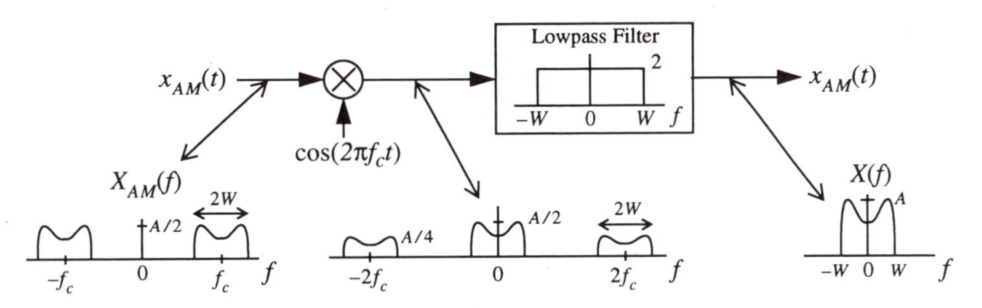

The process seems quite simple, but there is a catch. The $\cos(2\pi f_c t)$ must match *exactly* both in frequency and in phase with the cosine used to modulate the signal. In practice, this can be accomplished in the receiver by using a device called a phase-locked loop (PLL) to "lock on" to the carrier signal present, thus internally generating a suitably matched demodulating carrier signal. See page 511 in Section 17.1 of *Circuits, Signals, and Systems* by Siebert (1986) for a discussion of the effects of demodulating with a carrier of mismatched phase.

AM with Carrier Modulation and Envelope Detection

In the early days of radio, building a synchronous detector with a phase-locked loop was not a simple task. The parts were not available or they were simply too expensive. It was discovered that a slight change to the modulation process would make a much simpler detector possible.

That change is called AM with carrier (AM-WC) transmission. The only difference between AM-WC and standard AM is the addition of a constant (DC offset) to the original signal. The modulating signal then becomes $x_c(t) = x(t) + C$ instead of just $x(t)$, which adds an impulse at the origin in the spectrum of the input signal.

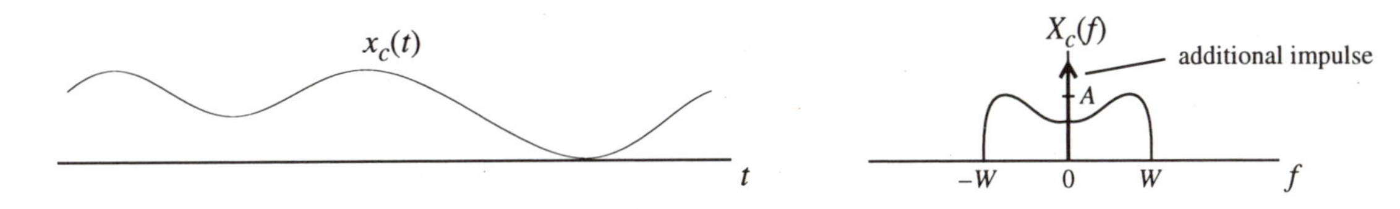

After multiplying by the cosine carrier signal, $x_{wc}(t)$ in the time and frequency domains looks like:

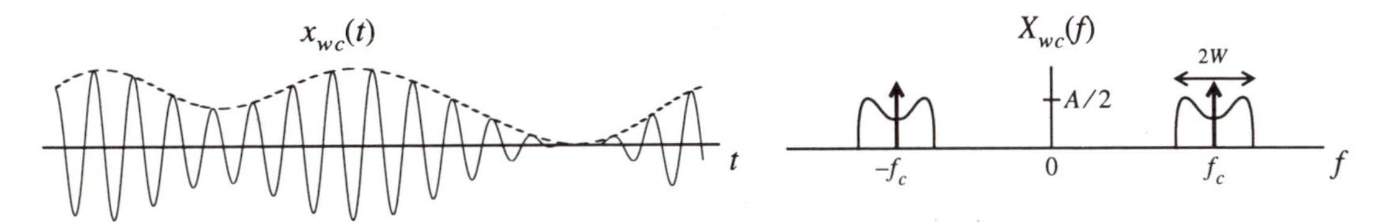

Given that the carrier frequency is high enough, notice that the upper envelope of an AM-WC modulated signal is equal to the original input signal $x_c(t)$. Since we know that the signal $x_c(t)$ is always greater than zero, we can use a "peak detector" circuit to track the envelope of $x_{wc}(t)$ and produce an approximation of $x_c(t)$ at its output. A simple peak detector is shown below. It consists of essentially just three parts and costs only pennies.

Peak Detector Circuit

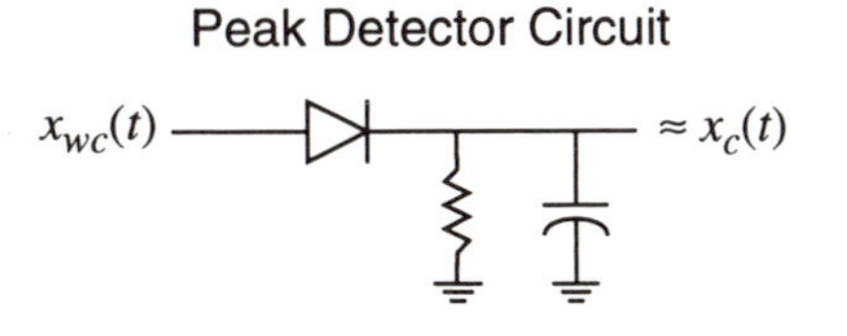

Naturally, there must be some tradeoffs associated with having such a simple detector. One is apparent from the above diagram of $X_{wc}(f)$. There are two additional impulses that do not appear in the standard AM spectrum. These additional impulses can often have significant area. More area implies that more power (see Parseval's Theorem) is necessary to transmit AM-WC. A second tradeoff is a constraint that the new modulating signal $x_c(t) = x(t) + C$ must be greater than zero for all t. This means that a sufficiently large enough value of C must be chosen to handle all types of input signals $x(t)$. However, the larger C is, the more power is wasted in transmission. If $x_c(t)$ goes negative (since the chosen value of C was too small), the envelope of $x_{WC}(t)$ will not match the shape of $x(t)$, resulting in severe distortion in the peak detector output. This condition is referred to as "overmodulation."

Although modern AM radios have become a little more sophisticated, the standards for AM broadcasting remain essentially the same. Even today your favorite AM station still transmits an AM-WC signal. Now it should be apparent why AM radios are so cheap to buy.

Single Sideband Amplitude Modulation

Single sideband AM (AM-SSB) is a variation on standard AM designed to cut the bandwidth of the modulated signal in half. Some of the approaches for generating an AM-SSB signal are outlined in Section 17.3 of *Circuits, Signals, and Systems* by Siebert (1986) and Section 7.3 of *Signals and Systems* by Oppenheim et al (1983). In any case, if you just remember the basic properties of Fourier transforms and LTI systems, you should be able to break down and analyze any novel modulation scheme (a frequent type of exam question).

18.3 Frequency Modulation and Phase Modulation

Amplitude modulation systems use the modulating signal to vary the *amplitude* of a sinusoidal carrier signal. Another important class of transmission techniques is referred to as angle modulation, in which the modulating signal is used to control the *frequency* or *phase* of a sinusoidal carrier. These techniques are known as frequency modulation (FM) and phase modulation (PM) respectively. Two main advantages to these angle-based schemes are: (1) the transmitted signal is always at a fixed amplitude, thus eliminating fluctuations in transmitter power, and (2) any noise picked up during transmission usually shows up as amplitude variations or "fuzz" on the received signal; if this was an AM transmission, the extra noise would appear to be part of the signal. This is why, in general, FM radio stations sound clearer than AM stations.

Phase modulation is when the *phase* of the transmitted carrier signal varies in proportion to the input signal $x(t)$. Frequency modulation is when the *frequency* of the transmitted signal transmitted varies in proportion to the input signal $x(t)$. Note that the instantaneous frequency of a sinusoid whose frequency is changing in time is the derivative of the phase function. Thus, phase modulation and frequency modulation are closely related. Phase modulating with $x(t)$ is identical to frequency modulating with the derivative of $x(t)$. These results are summarized below:

Phase Modulation

$$x_{PM}(t) = A\cos(2\pi f_c t + \theta_c(t))$$

$$\text{where } \theta_c(t) = \theta_0 + k_p x(t)$$

The phase $\theta_c(t)$ of the carrier sinusoid (frequency f_c) varies around a central value θ_0 in proportion to the input signal $x(t)$.

Frequency Modulation

$$x_{FM}(t) = A\cos\phi(t)$$

$$\text{where } \frac{d\phi(t)}{dt} = 2\pi f_c + k_f x(t)$$

The frequency of $x_{FM}(t)$ varies around a central carrier frequency f_c in proportion to the input signal $x(t)$. For example, if $x(t)$ is a constant, then $x_{FM}(t)$ is a constant frequency sinusoid.

$\frac{d\phi(t)}{dt}$ is known as the *instantaneous frequency*

An example of frequency modulation is shown below.

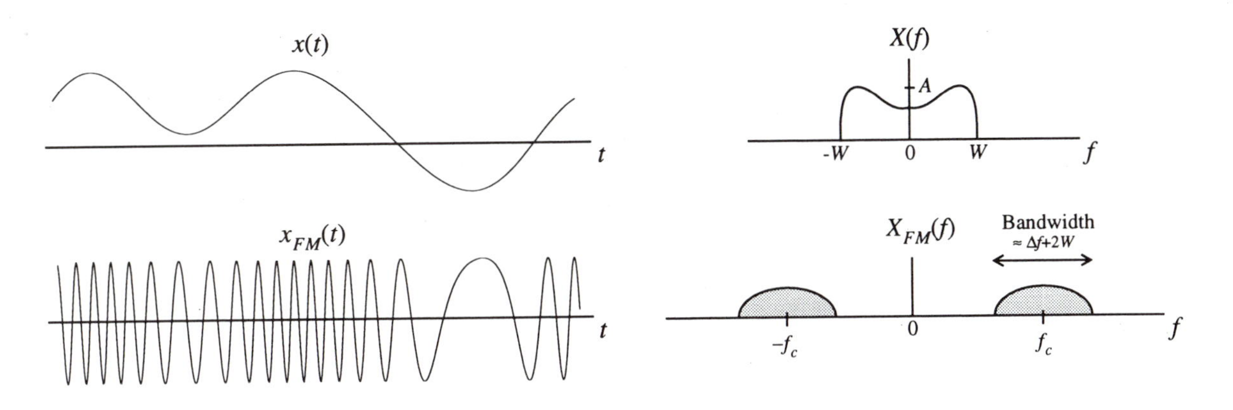

Notice that the "instantaneous frequency" of $x_{FM}(t)$ is proportional to the amplitude of $x(t)$. The exact spectrum of an FM modulated signal is very difficult to determine; in general, it looks nothing like the shape of the input signal's spectrum. $X_{FM}(f)$ shown in the above diagram is presented only to depict the location and approximate bandwidth of the FM signal.

FM Bandwidth (Wideband & Narrowband FM)

As mentioned above, the Fourier transform $X_{FM}(f)$ is not easily found. However, it is important to have some idea of its bandwidth so we know how close adjacent signals can be placed when frequency multiplexing. Let's define $\Delta f = f_{xmax} - f_{xmin}$ as the "frequency deviation." The values f_{xmax} and f_{xmin} correspond to the output frequency if a constant $x(t) = xmax$ or $x(t) = xmin$ is used as the input signal. We can then approximate the bandwidth of $X_{FM}(f)$ using the following expression:

$$\text{Bandwidth} \approx \Delta f + 2W$$

$$\text{Wideband} \Rightarrow \Delta f \gg 2W \Rightarrow \text{Bandwidth} \approx \Delta f$$

$$\text{Narrowband} \Rightarrow 2W \gg \Delta f \Rightarrow \text{Bandwidth} \approx 2W$$

The value $2W$ is the bandwidth of the input signal $X(f)$. There are two types of frequency modulation: wideband and narrowband. Each refers to the amount of frequency fluctuation in the modulated signal, which essentially corresponds to the size of the constant k_f in the FM equations described earlier. The benefit of using a wide bandwidth with FM is improved noise rejection; however, the signal then occupies more space in the frequency world. Once again, we have a tradeoff.

FM Demodulation

An FM signal can be demodulated to recover the original signal $x(t)$ through the use of a phase-locked loop. This device locks on to the phase of the input signal; differences between the expected phase and the actual phase (frequency) of the incoming sinusoid result in an output error signal from the device. This error signal varies in exactly the same manner as the changes in frequency during the modulation process; thus it is directly proportional to the original signal $x(t)$.

18.4 Superheterodyne Receivers

The function of a radio receiver can be broken down into two steps: (1) tune its filters to focus on the desired frequency band from a frequency multiplexed channel, and (2) recover (demodulate) the received signal. The superheterodyne receiver is a method of efficiently achieving the first stage in the above process. It is found in virtually all of today's radios. One way to build a receiver is to use a sharp cutoff bandpass filter that can be tuned over a range of several megahertz. However, such filters are expensive and difficult to build. The basic idea behind the superheterodyne scheme is to build a high quality stationary filter (much cheaper) and move the signal to the filter instead of having to move the filter to the signal. The basic steps are as follows:

1. Use a poor quality tunable bandpass filter (BPF) to initially focus on the desired station.
2. Frequency shift the spectrum to a fixed intermediate frequency (IF).
3. Apply a sharp, carefully designed BPF centered at the IF to remove any remaining stray spectra.
4. Proceed with standard demodulation scheme (i.e. synchronous detector).

Again, instead of moving your sharp bandpass filter to the station, the superheterodyne scheme moves the station to your sharp BPF. For example, let's assume that three separate signals have been AM modulated using three distinct carrier frequencies. The spectrum of the transmitted signal might look something like this:

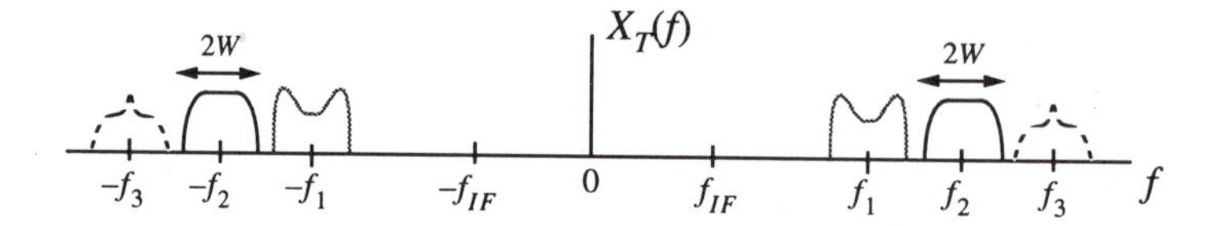

For this example, we would like to demodulate the signal located at frequency f_2 using a superheterodyne system. A block diagram of such a system is given below.

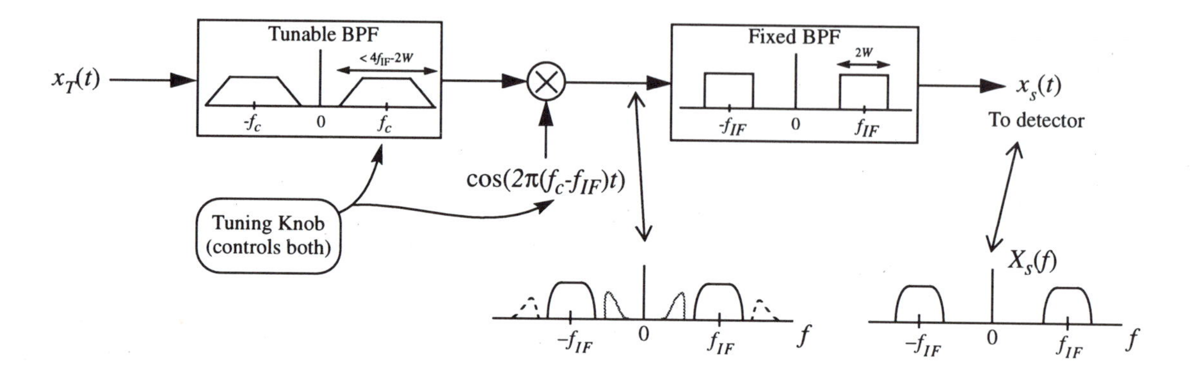

Notice that two bandpass filters are used. The first BPF has a wide bandwidth (but must be less than $4f_{IF} - 2W$ where W=maximum frequency in signal) and has a tunable center frequency that changes in conjunction with the frequency of the cosine wave. This type of dual tuning allows the isolation of a single frequency range. In our case, tuning $f_c = f_2$ selects the signal we want.

The multiplication by a $\cos(2\pi[f_c-f_{IF}]t)$ shifts the desired spectrum to f_{IF} where the second bandpass filter does the final clean up job. The result is the spectrum $X_s(f)$, which is now ready for the synchronous detector to complete the final shifting to the audible frequency range (centered about zero).

18.5 A Sample Problem

Question:

An input signal $x(t)$ has a Fourier transform $X(f)$ whose real and imaginary parts are shown below.

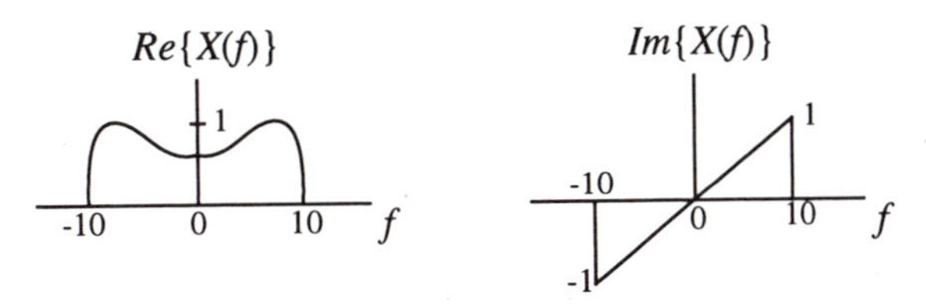

The signal $x(t)$ is multiplied by $\sin(2\pi 50t)$ and $\cos(2\pi 50t)$ as shown here:

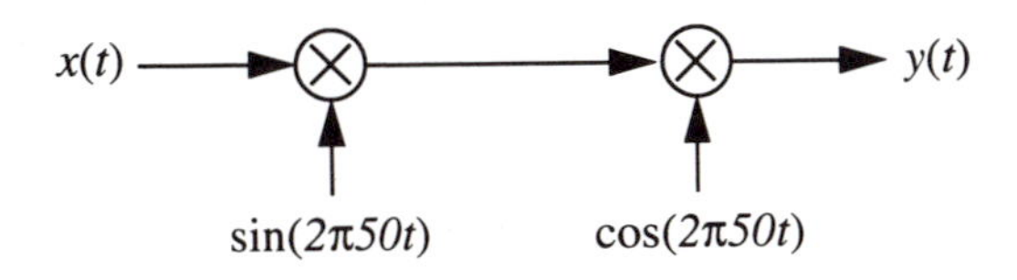

Sketch $Y(f)$.

Answer:

A shortcut is to use the identity, $2\sin\theta\cos\theta = \sin 2\theta$, but the long way isn't hard either. Watch for sign errors!

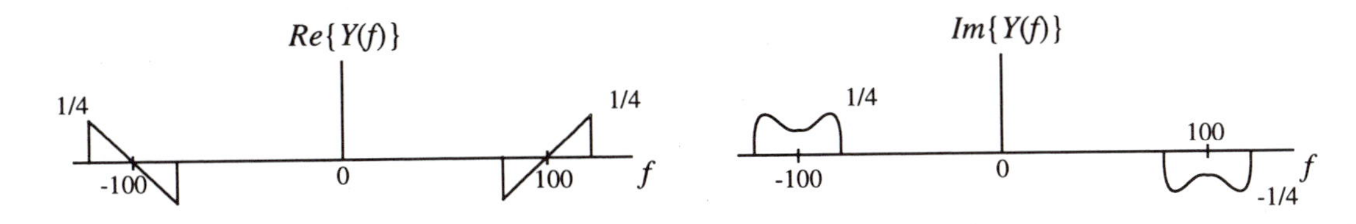

CHAPTER 19 Sampling

Overview

Sampling theory has become more and more critical in the "digital age." Most signals that we work with are continuous, but they are increasingly being processed using digital computers. Sampling theory gives us the tools to faithfully convert from the analog world to digital and back again. The fundamental result of this chapter is that any bandlimited signal can be *completely* characterized by discrete, equally spaced samples, provided that the samples are spaced close enough together in time.

19.1 What is Sampling?

Sampling is the process of taking a continuous-time signal and representing it by a series of discrete samples. Reconstruction is the process of taking these discrete samples and recreating the associated continuous-time signal. These processes are illustrated in the following block diagram.

Overall process:

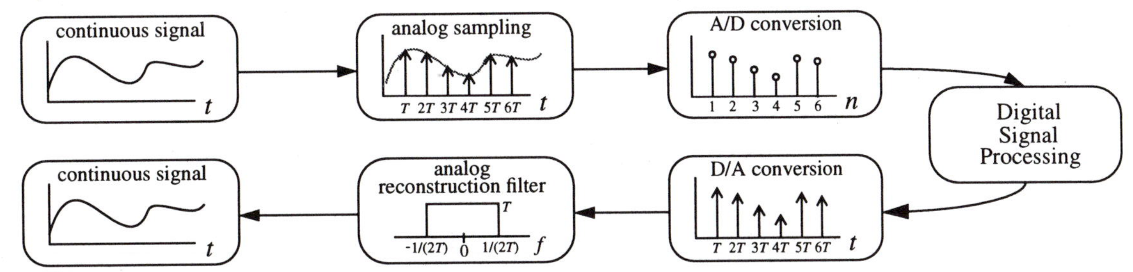

We will leave the processes of A/D, D/A, and digital signal processing for another book. Why do we need sampling theory at all? We could just work with analog signals only. Doing signal processing in the digital world, i.e. on computers, provides far more flexibility, but with the expense of added complexity. Filter characteristics are easily changed by reprogramming a few numbers instead of having to change resistors and capacitors. Also, the processing is essentially noise-free, unlike the "fuzz" that you sometimes see on your oscilloscopes when doing analog circuit design.

This book:

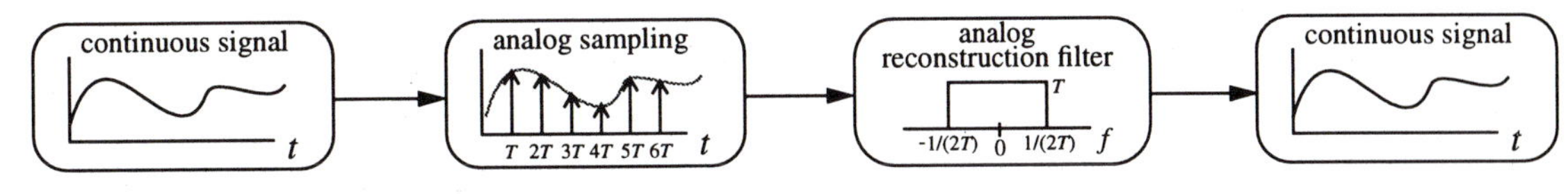

19.2 Mechanisms of Sampling

Time Domain

Let's examine the sampling process in closer detail, starting in the time domain. The standard paradigm is shown in the following picture. The value of the continuous-time signal is recorded every T seconds by multiplying by an impulse train. The area of the pulses formed in the sampled version is equal to the height of original signal at the sample point.

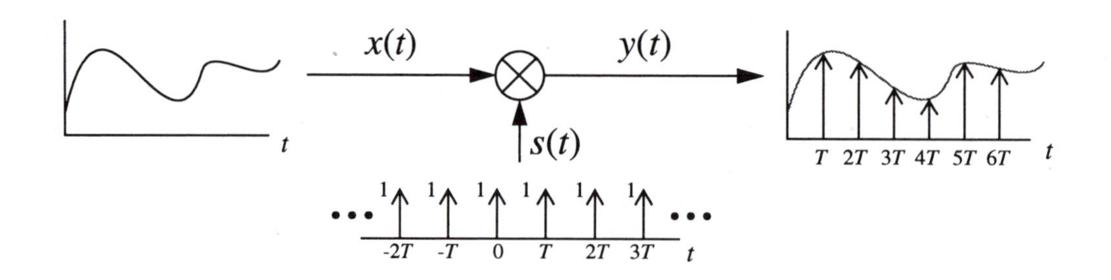

Frequency Domain

Now, let's look at the same process in the frequency domain. First, we assume a bandlimited spectrum for $X(f)$, the Fourier transform of $x(t)$. Remember, a bandlimited spectrum is one where $X(f)=0$ for $|f| > W$. The reasons for needing a bandlimited spectrum will become clear shortly. The following three facts allows us to determine the spectrum for $Y(f)$, as graphed below.

1. The transform of an impulse train is an impulse train.
2. Multiplication in the time domain becomes convolution in the frequency domain.
3. Convolving with an impulse merely shifts and scales the signal.

Do not forget to take into account the scale in height of $1/T$ that accompanies the transform of the impulse train.

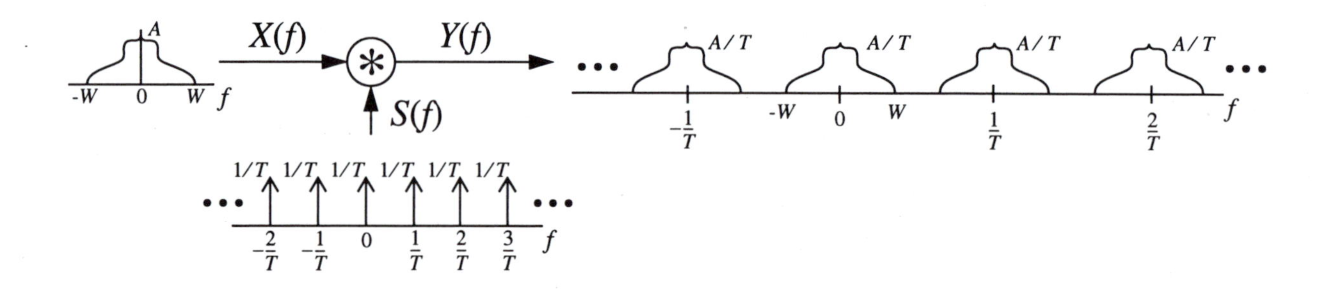

19.3 The Reconstruction Process

Frequency Domain

The original continuous-time signal $x(t)$ is completely recovered if the original spectrum $X(f)$ can be extracted from $Y(f)$, the spectrum of the sampled signal. By examining the picture of $Y(f)$ shown below, it should be clear that the single spectrum $X(f)$ can be recovered by using an ideal lowpass filter to remove the extra copies. The filter should have cutoffs at $f = \pm 1/(2T)$ and a gain of T.

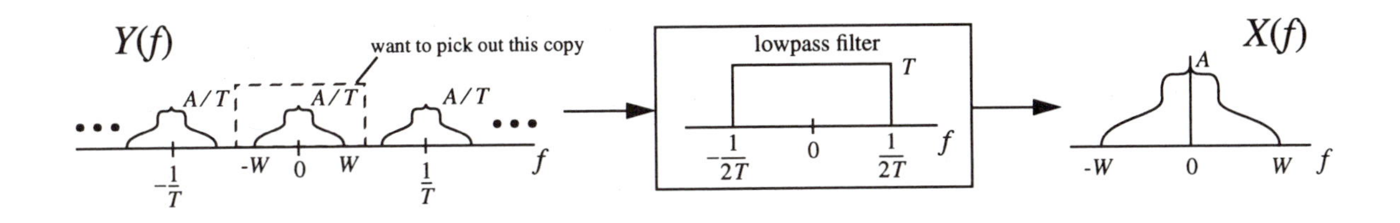

Time Domain

To see how the reconstruction process works in the time domain, recall that a lowpass filter in the frequency domain is a sinc function in the time domain. So, multiplying by a "box" filter in frequency is like convolving with a sinc function in time. Since $y(t)$ is just a series of impulses (samples of $x(t)$), it follows that the reconstructed $x(t)$ is merely the superposition of scaled and shifted sincs!

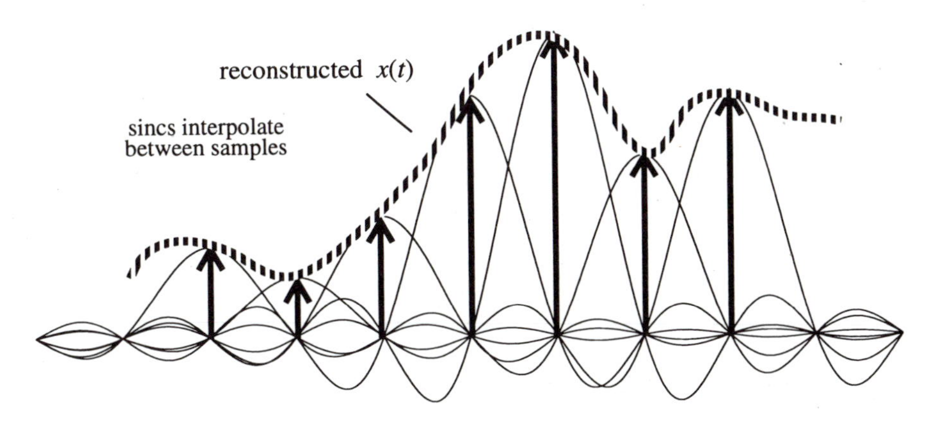

19.4 Aliasing

We have seen that a continuous-time signal can be completely reconstructed from its samples. However, this process will not be possible if the samples are spaced too far apart in time. Why? As the impulses in the impulse train in time are spaced farther and farther apart, the impulses in its transform $S(f)$ get closer and closer together. If the impulses in frequency move too close together, after convolution the copies of the spectra of

$X(f)$ will overlap. Bad! Once the spectrum gets garbled like that, it will be impossible to recover the original continuous-time signal. The situation where spectral overlap occurs is known as *aliasing*.

High frequency information from the adjoining spectrum moves into the low frequency range of the main spectrum, implying that high frequencies are being aliased into looking like lower frequencies.

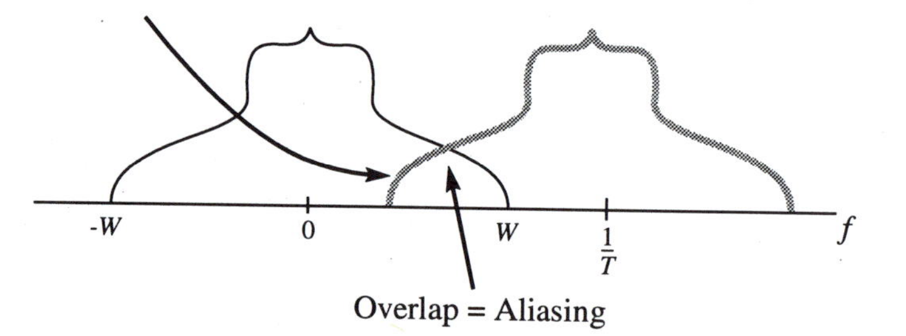

Overlap = Aliasing

In the case of overlap, the high frequency information aliases/resembles/looks-like low frequency information. This effect is demonstrated below with a high frequency sine wave that is sampled too slowly. Even though the sine wave is sampled at regular intervals, the reconstruction process fails to reproduce the original signal since the samples are spaced too far apart in time. It is possible to fit a lower frequency sine wave to those same data points, which is a time-domain illustration of how a high frequency signal can get aliased into looking like a lower frequency signal.

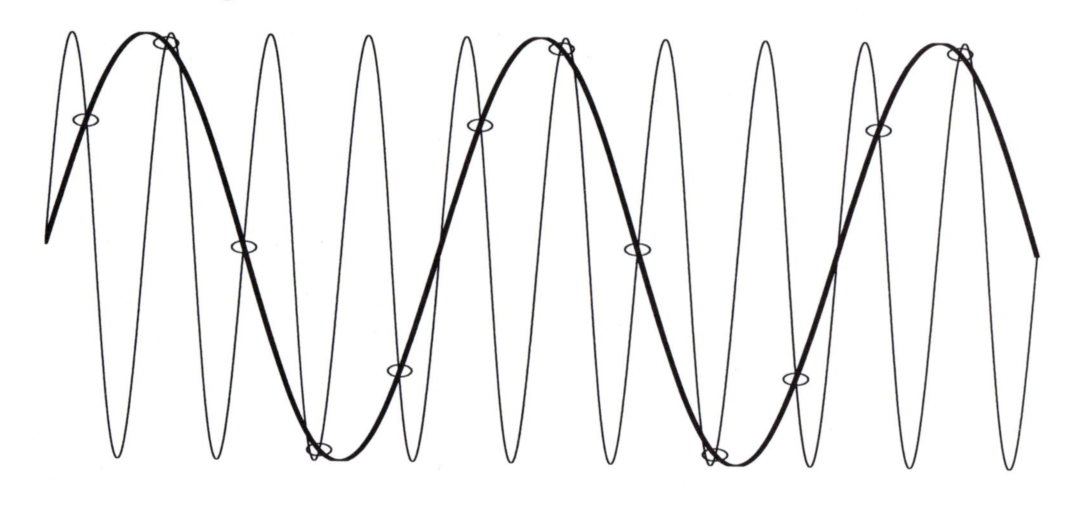

19.5 The Nyquist Sampling Theorem

In order to have distortion-free reconstruction of a continuous-time signal, two things have to happen: (1) the continuous-time signal to be sampled must be bandlimited, and (2) the samples must be close enough together in time. From examining the spectrum for $Y(f)$ in Section 19.3, to prevent overlap we must have $1/T - W > W$, or $1/T > 2W$. Since T is known as the sampling period (in seconds), $1/T$ is called the sampling rate (in Hertz). All of this implies the grand result that in order to achieve flawless theoretical reconstruction:

> *The sampling rate must be greater than twice the maximum frequency present in the input signal.*

The critical sampling rate ($2W$) is known as the Nyquist rate. The above statement is known as the Nyquist Sampling Theorem and is definitely one of the most powerful concepts ever seen by an engineering student.

19.6 Practical Considerations

Zero-Order Hold

However nice theory may seem, impulses don't really exist in the real world. In practice, what is commonly done for sampling purposes is known as the zero-order hold. Here, the sampled signal resembles a staircase (i.e. remains flat between sample points). It is still possible to completely reconstruct the original signal; however, the lowpass filter must be slightly modified. See Section 8.1.2 of *Signals and Systems* by Oppenheim et al (1983) for more information. Note that using a standard flat lowpass filter for reconstruction would probably work reasonably well, but it wouldn't be optimal.

Anti-Aliasing Filters

To ensure that aliasing does not occur in the sampling process, people often first pass the signal to be sampled through a lowpass filter in order to insure that it is appropriately bandlimited. This filter is known as an anti-aliasing filter. For example, if you want to sample audio at a relatively low rate of 5KHz, you would first need to pass the sound source through a lowpass filter that had a sharp cutoff at 2.5KHz or lower in order to avoid aliasing of any higher frequency content present in the original signal.

Oversampling

The Nyquist theorem states that it is possible to completely reconstruct a signal as long as you sample at a rate greater than twice the maximum frequency present in the signal. From Section 19.3, if $1/T = 2W$ then the copies of the spectra in $Y(f)$ will be adjacent, i.e. touching. So, in order to reconstruct the original signal you are going to need a very, very sharp analog lowpass filter. Such beasts are not easy to design, so what is commonly done is to *oversample* – that is, to sample at a rate much greater than the Nyquist rate. Then, the copies of the spectra are spread sufficiently far apart so that a more realistic lowpass filter can be used in the reconstruction process. This is why some compact disc players are advertised as having 4x oversampling, 8x oversampling, etc.

A Little Trivia

- The human auditory system can only hear sounds between 20 Hz and 20 KHz.
- Compact discs contain audio sampled at a rate of 44.1 KHz.
- The bandwidth of an ordinary telephone line is about 3KHz (higher frequencies are lost during transmission), which is still large enough to produce an acceptable quality of conversational speech.

19.7 A Sample Problem (no pun intended)

Questions:

Using the block diagram shown below, answer the following questions:

(a) Sketch $X_s(f)$ for T=20 msec. The real and imaginary parts of the spectrum $X(f)$ are given.

(b) What is the maximum value of T for which the system will produce $y(t) = Kx(t)$ (i.e. perfect reconstruction within a constant factor)?

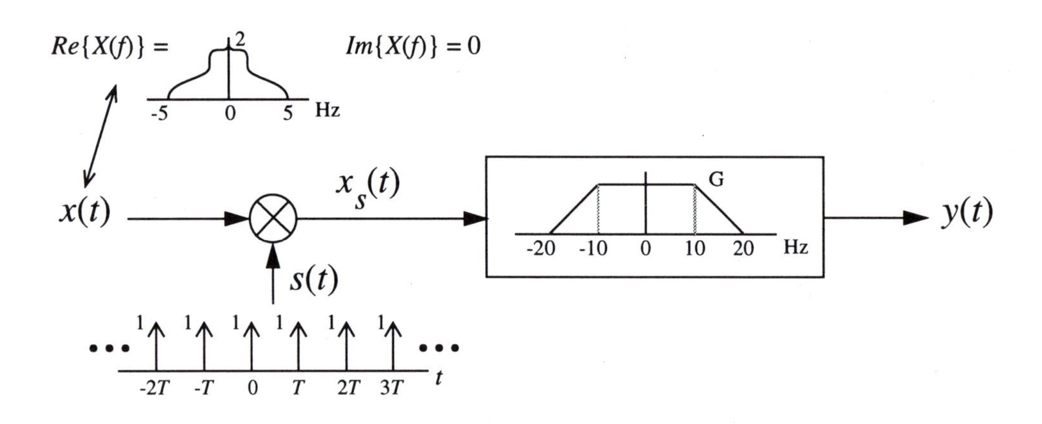

Answers:

(a) Copies of the spectra of $X(f)$ are replicated at -100, -50, 0, 50, 100, etc. Hertz. Height of each is 100.

(b) Maximum value of T is 40 msec. Cannot sample any slower because of the limitations of the lowpass filter.

CHAPTER 20 Fourier Series

Overview

The Fourier transform of an aperiodic signal produced a continuous transform $X(f)$. When the input signal is periodic, the Fourier transform is known as the Fourier series. The reason for calling it a series is that instead of needing a continuous function to describe the spectrum, the frequency content of a periodic signal can be represented by a discrete set of numbers. These numbers are called the Fourier series coefficients. They provide the weights on the harmonically related sinusoids that can be used to reconstruct the original signal. The values of these coefficients can be found through direct computation or by sampling the Fourier transform of just one period of the input signal.

20.1 Orthogonal Basis Functions

It is well known that practically any function can be written as the weighted sum of a set of orthogonal functions. This idea is similar to the idea of representing a vector as the weighted sum of unit vectors. There are several different possible orthogonal basis function sets, but Dr. Fourier decided to use sinusoids. The Fourier transform simply illustrates the weights on the various sinusoids needed to reproduce the input signal.

20.2 What is the Fourier Series?

Periodic functions are completely described by the weights on a *discrete* set of sinusoidal frequencies, as opposed to a continuous range of frequencies, as with the Fourier transform for aperiodic functions. This set of weights is known as the Fourier series coefficients. Furthermore, the sinusoids that go with these coefficients are harmonically related; each one's frequency is an integer multiple of the fundamental frequency of the input signal. The fundamental frequency is the reciprocal of the length of one period of the input signal.

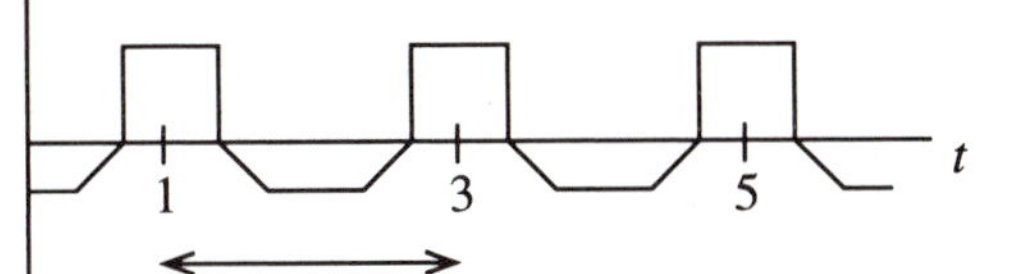

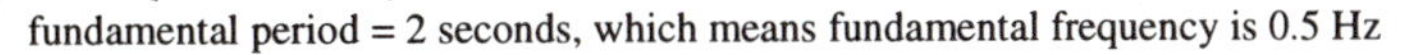

fundamental period = 2 seconds, which means fundamental frequency is 0.5 Hz

This periodic signal can be completely described using sinusoids of frequencies 0, 0.5, 1, 1.5, 2, 2.5, etc. Hertz.

20.3 Forms of the Fourier Series

Sine-Cosine Form

Any periodic function $x(t)$ can be written as the sum of harmonically related sinusoids in the following manner. Note that T denotes the length in seconds of the fundamental period of the signal. There are additional formulas for determining the value of the coefficients a_n and b_n, but there is a more intuitive way of obtaining these values, as we will see shortly.

$$x(t) = a_0 + \sum_{n=1}^{\infty} a_n \cos\frac{2\pi nt}{T} + \sum_{n=1}^{\infty} b_n \sin\frac{2\pi nt}{T}$$

Magnitude Phase Form

Using the fact that two sinusoids of the same frequency can be combined into a single sinusoid with a change of magnitude and phase, the cosine and sine terms in the above expression can be combined and written as:

$$x(t) = a_0 + \sum_{n=1}^{\infty} c_n \cos\left(\frac{2\pi nt}{T} + \theta_n\right)$$

Complex Exponential Form

However, the simplest and most useful form is the exponential form given by:

$$x(t) = \sum_{n=-\infty}^{\infty} X[n]\, e^{jn2\pi ft} \qquad X[n] = \frac{1}{T}\int_{period} x(t)e^{-jn2\pi ft}dt$$

$$f = 1/period$$

In general, $X[n]$ is a complex number

Relationship among Forms of the Fourier Series

The coefficients in the various forms of the Fourier series are all interrelated as shown below:

$X[n]$ is a complex number

$X[n] = \frac{1}{2}(a_n - jb_n)$

$X[n] = \frac{1}{2}c_n e^{j\theta_n}$

$a_0 = X[0]$

$a_n = 2Re\{X[n]\}$

$b_n = -2Im\{X[n]\}$

$c_n = \sqrt{a_n^2 + b_n^2}$

$\theta_n = -\tan^{-1}(b_n/a_n)$

20.4 Relationship Between the Fourier Series and Transform

Note that any non-periodic signal can be considered to be a periodic signal with period set to infinity. Doing this will cause the Fourier series formulas to collapse to the Fourier transform formulas. For more information, read Section 4.4 of *Signals and Systems* by Oppenheim et al (1983).

The Fourier series coefficients of a periodic signal are proportional to equally spaced samples of the Fourier transform of one period of the input signal. To see this, first convince yourself that any periodic signal can be represented as a single period convolved in time with an impulse train as shown below:

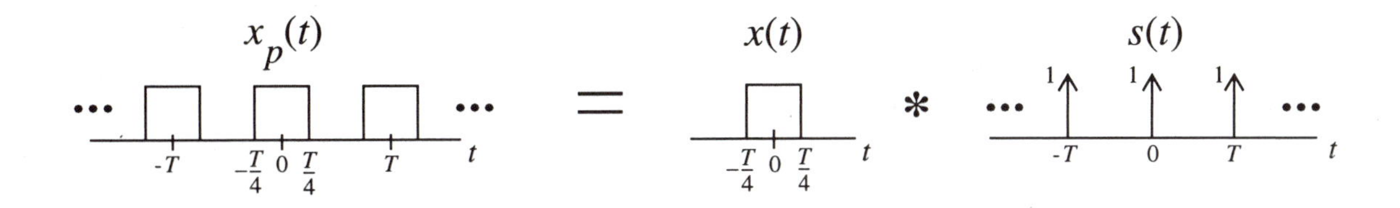

So, to find the transform of $x_p(t)$, we could just find the transform of $x(t)$ and *multiply* it by the transform of $s(t)$ since convolution in the time domain equals multiplication in the frequency domain. Using the above example, we can generate the following frequency domain picture:

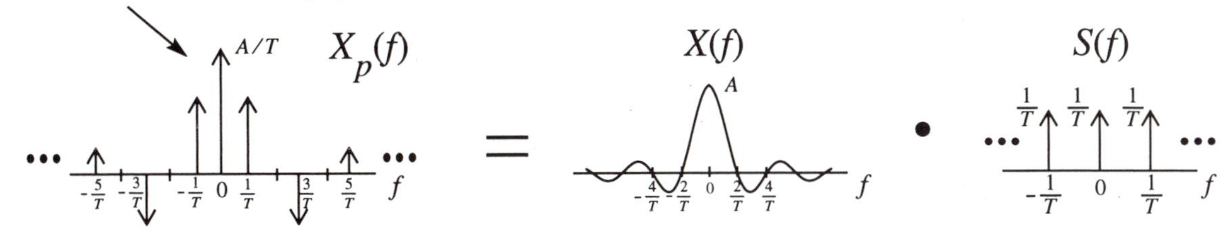

Summary

The Fourier transform of any *periodic* signal is <u>always</u> composed of just impulses. The area of these impulses are the Fourier series coefficients for the exponential form. Furthermore, these coefficients can be obtained by sampling the Fourier transform of one period of the signal at frequencies which are multiples of $1/T$.

20.5 The DC Offset

Note that $X[0]$ or a_0, the first term in the Fourier series, is merely the average value or DC offset of the periodic signal. Those signals that have equal "weight" above and below the horizontal axis have no DC offset, and hence are equal to zero at the origin in the Fourier transform and series.

20.6 Properties of the Fourier Series Coefficients

There are several special cases of periodic input signals that immediately imply certain properties of the Fourier series coefficients. Some of these rules will seem similar to those of the Fourier transform since the Fourier series is just a special case of the transform. For the table below, assume $x(t)$ is real.

Description	*Condition*	*$X[n]$ coeffs*	*a_n, b_n coeffs*
even	$x(t) = x(-t)$	purely real	all $b_n = 0$
odd	$x(t) = -x(-t)$	purely imaginary	all $a_n = 0$
odd harmonic	$x(t - T/2) = -x(t)$	cmplx, $X[n]=0$, n even	$a_n \& b_n = 0$, n even
even harmonic	$x(t - T/2) = x(t)$	cmplx, $X[n]=0$, n odd	$a_n \& b_n = 0$, n odd

Notes:

- a_0, the DC offset term, can be non-zero even though all the other a_n's are zero.
- An odd-harmonic function is one where the second half of its period is the negative of the first half.
- An even-harmonic function is one where the second half of its period is exactly the same as the first half. Therefore, any function that is even-harmonic is actually a regular periodic function whose period has been labeled twice what it should be. In other words, there is nothing special about even-harmonic functions.
- Shifting a signal left/right in time does not affect whether or not it is odd-harmonic.
- Shifting a signal up/down (adding a DC offset) does not affect whether it is odd-harmonic, other than adding a term in the Fourier series at zero frequency.
- An odd-harmonic function does not have to be odd.

Some Examples:

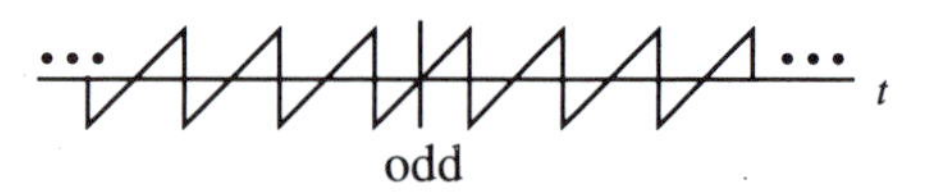

odd

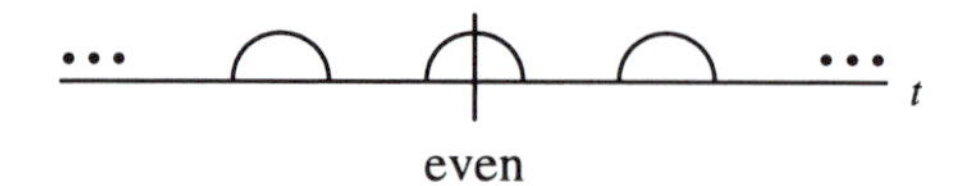

even

odd-harmonic

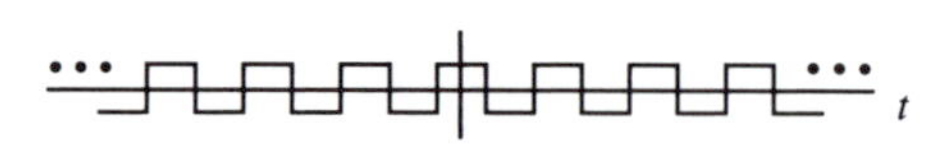

even and odd-harmonic

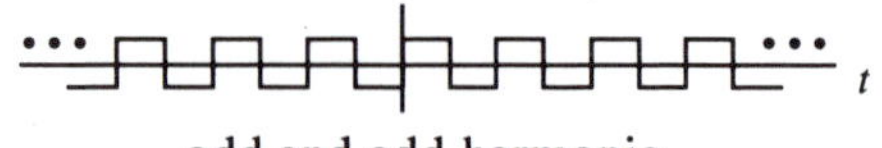
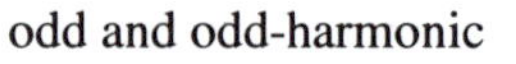

odd and odd-harmonic

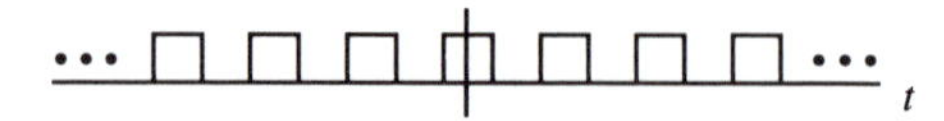

even and odd-harmonic w/ DC offset

20.7 Parseval's Theorem for Fourier Series

Parseval's theorem equates the average "energy" in one period of the input signal with its "energy" in the frequency domain.

$$\frac{1}{T}\int_T |x(t)|^2 dt = \sum_{n=-\infty}^{\infty} |X[n]|^2$$

20.8 Square Wave Reconstruction

As an exercise to test your understanding of the material, verify that the following square wave can be represented by the cosine series shown. The approach is to first find the Fourier transform of the periodic signal by representing the square wave as a single pulse convolved with an impulse train. The transform will be then be a sinc function multiplied by an impulse train in frequency. This product produces a series of impulses whose areas are the scaled samples of the sinc function. Read Section 20.4 again if this didn't make any sense. The areas of these impulses are the Fourier series coefficients $X[n]$ for the exponential form (see Section 20.3). Each symmetric pair of impulses represents a cosine wave in the time domain.

$x(t) =$

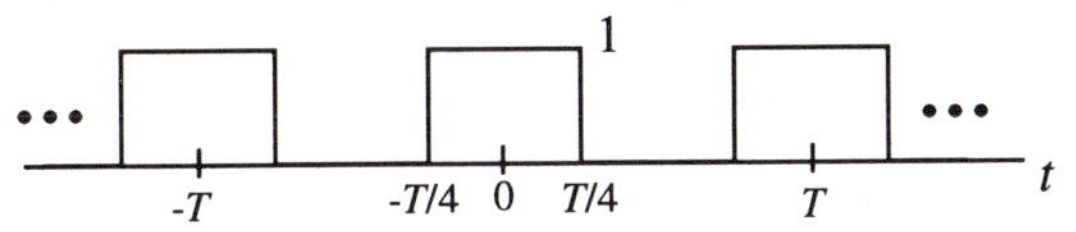

$$x(t) = \frac{1}{2} + \sum_{k \text{ odd}} \frac{2(-1)^{(k-1)/2}}{\pi k} \cos\frac{2\pi k t}{T}$$

We have just shown that the sum of cosine waves can produce a square wave! That probably doesn't seem obvious, so let's demonstrate it visually. The following plots show the sum of the first 1, 2, 5, and 20 terms in the above formula.

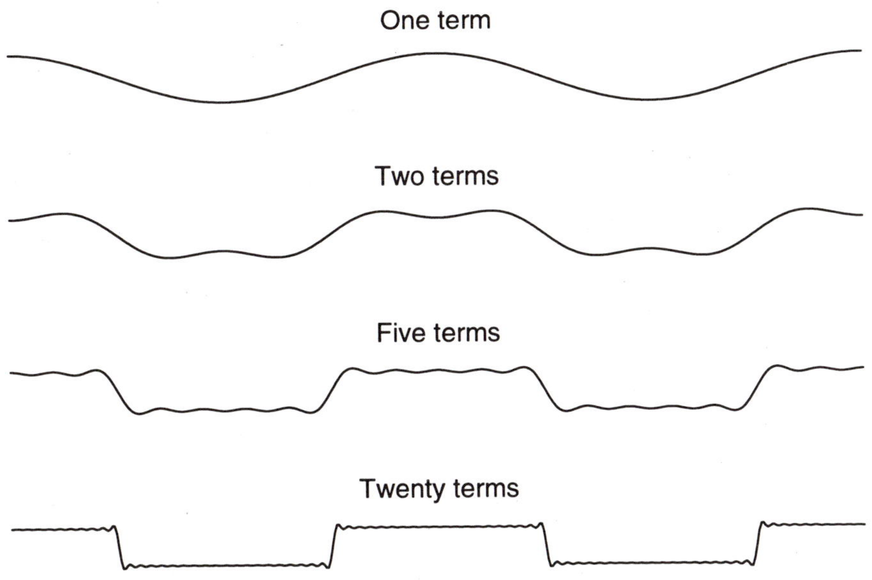

The Gibbs Phenomenon

If you look carefully, you will notice that there are some ripples near the transitions of the reconstructed square wave. Increasing the number of terms in the summation increases the frequency of their oscillation, but does not decrease their amplitude. In fact, the peak overshoot remains fixed at about 9% of the height of the discontinuity (height of square wave); however, the total *energy* in the oscillations does decrease as more terms are used. This behavior is known as the Gibbs phenomenon. See Section 15.3 of *Circuits, Signals, and Systems* by Siebert (1986) and Example 4.6 on page 184 of *Signals and Systems* by Oppenheim et al (1983) for more information.

20.9 A Sample Problem

A periodic square wave signal is passed through a filter whose impulse response is shown below. Find the equation for the output signal $y(t)$.

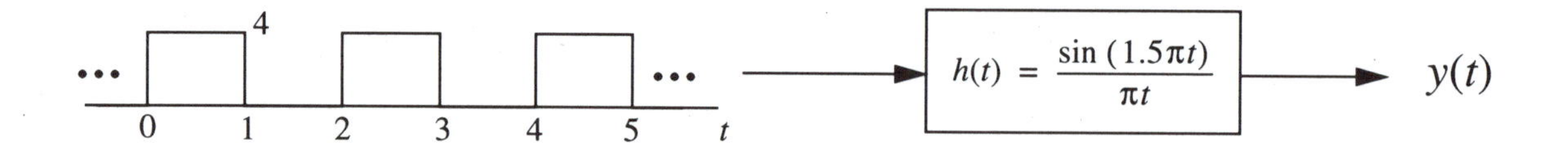

Answer:

This problem is a bit tricky. Notice that the square wave as shown is not an even function; this means that its Fourier transform is not purely real, which makes for a bit of a mess. An easy way to get around this is to make the input even by shifting it 0.5 seconds to the left. After finding the output for this modified input signal, just shift the answer to the right by 0.5 seconds to get the true output. Remember, we are allowed to perform such manipulations since this is a time-invariant system.

The system designated by $h(t)$ is an ideal lowpass filter that has cutoffs at $f = \pm 3/4$Hz. Only three terms in the Fourier transform of the periodic square wave make it through the filter: an impulse at the origin and two impulses at $f = \pm 1/2$. Taking the inverse transform, these three impulses correspond to a cosine wave with a DC offset. If you didn't get the right answer, be sure you took into account the height of the input square wave as well as the $1/T$ scaling factor on the transform of an impulse train.

$$y(t) = 2 + \frac{8}{\pi}\cos(\pi t)$$ (answer if input is shifted 0.5 seconds to the left)

$$y(t) = 2 + \frac{8}{\pi}\cos(\pi(t - 0.5))$$ (shift the answer 0.5 seconds back to the right)

$$y(t) = 2 + \frac{8}{\pi}\cos(\pi t - \pi/2)$$

$$y(t) = 2 + \frac{8}{\pi}\sin(\pi t)$$

APPENDIX Review Topics

Overview

This chapter is designed to review a few basic mathematical issues you will likely encounter in the study of signals and systems. Topics covered are complex numbers, linear circuit theory, the quadratic formula, trigonometric identities, partial fraction expansion, logarithms, sequences and series, binomial expansions, and linear algebra.

A.1 Complex Numbers

What is a complex number? What is an imaginary number? Why do we need them? We all know about real numbers – examples include 0, 1, -5.2, 3.1415926, $\sqrt{57}$, etc. In order to describe the square root of negative real numbers, the "number" j was created (electrical engineers will use j instead of i since i is primarily used to represent current). It is defined as the principal square root of -1.

$$\boxed{j = \sqrt{-1}}$$

For example, $\sqrt{-49} = \sqrt{(-1)(49)} = \sqrt{49}\sqrt{-1} = 7j$. Any number with a j in it is known as an *imaginary* number. A *complex* number is a number consisting of both a real and imaginary part.

$$\boxed{C = a + bj = \text{complex number}}$$

Some example complex numbers are $4 + 2j$ and $-2.3 - \sqrt{5}j$. An alternative way to express a complex number is by talking about its *magnitude* and *phase*. This is best seen by drawing the complex number as a vector in the complex plane. It is also common to refer to the real and imaginary parts of a complex number separately.

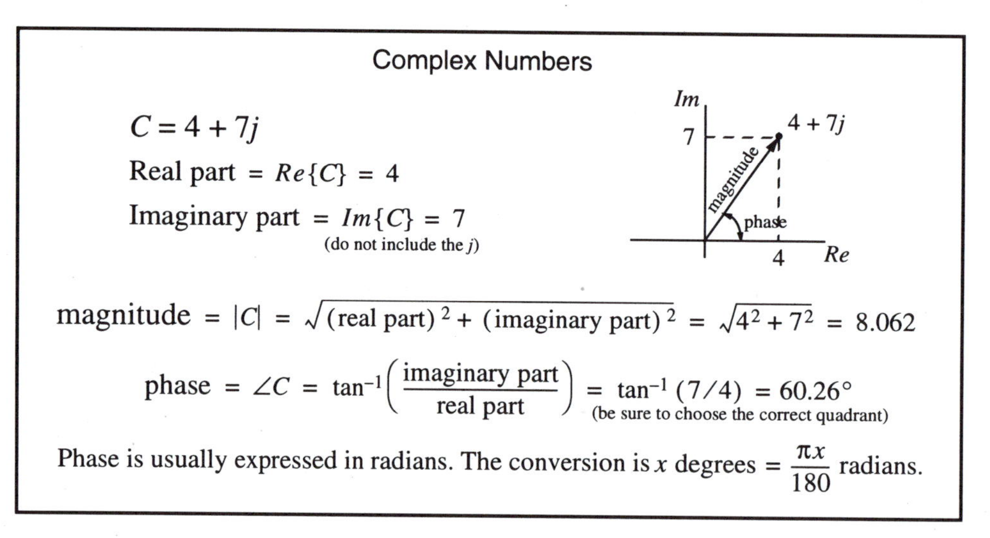

Complex Numbers

$C = 4 + 7j$

Real part $= Re\{C\} = 4$

Imaginary part $= Im\{C\} = 7$
(do not include the j)

$$\text{magnitude} = |C| = \sqrt{(\text{real part})^2 + (\text{imaginary part})^2} = \sqrt{4^2 + 7^2} = 8.062$$

$$\text{phase} = \angle C = \tan^{-1}\left(\frac{\text{imaginary part}}{\text{real part}}\right) = \tan^{-1}(7/4) = 60.26°$$
(be sure to choose the correct quadrant)

Phase is usually expressed in radians. The conversion is x degrees $= \frac{\pi x}{180}$ radians.

Euler's identity is a fundamental result in the area of complex number analysis. It can be verified using the sine, cosine, and exponential power series described in Section A.7.

Euler's Identity and Relations

$$e^{j\theta} = \cos\theta + j\sin\theta$$

$$\cos\theta = \frac{e^{j\theta} + e^{-j\theta}}{2} \qquad \sin\theta = \frac{e^{j\theta} - e^{-j\theta}}{2j}$$

The vector diagram for a complex number as shown above together with Euler's formula lead us to the compact representation of a complex number known as *polar form.*

Polar Form

$$C = a + bj = r(\cos\theta + j\sin\theta) = re^{j\theta}$$

r = magnitude $= \sqrt{a^2 + b^2}$ $\qquad$ θ = phase $= \tan^{-1}(b/a)$
(be sure to choose the correct quadrant)

$a = Re\{C\} = r\cos\theta$ $\qquad$ $b = Im\{C\} = r\sin\theta$

Operations on Complex Numbers

Addition/Subtraction: To add or subtract two complex numbers, merely add or subtract the real and imaginary parts separately and then combine. For example, $(2 + 3j) - (4 + j) = -2 + 2j$

Conjugation: The conjugate of a complex number C, denoted by C*, is obtained by negating the imaginary part of C. If $C = a + bj$ then $C^* = a - bj$. Note that the product CC^* = magntiude squared $= |C|^2 = a^2 + b^2$. Also note that, $(C_1C_2)^* = C_1^*C_2^*$ and $(C_1/C_2)^* = C_1^*/C_2^*$.

Multiplication: Multiply complex numbers just like you were multiplying two binomial expressions, keeping in mind that $j^2 = -1$. For example, $(2+3j)(4+j) = 8+2j+12j+3j^2 = 5+14j$.

Division: To divide two complex numbers, multiply both the numerator and the denominator by the complex conjugate of the denominator and simplify.

$$\frac{2+3j}{4+j} = \frac{(2+3j)(4-j)}{(4+j)(4-j)} = \frac{11+10j}{17} = \frac{11}{17}+\frac{10}{17}j$$

Multiplication and Division in Polar Form: It is much easier to perform multiplication and division when complex numbers are written in their polar form. To multiply two complex numbers in polar form, simply multiply the magnitudes and add their phases. When dividing, divide their magnitudes and subtract the phases.

$$(r_1 e^{j\theta_1})(r_2 e^{j\theta_2}) = r_1 r_2 e^{j(\theta_1+\theta_2)} \qquad \frac{r_1 e^{j\theta_1}}{r_2 e^{j\theta_2}} = \frac{r_1}{r_2} e^{j(\theta_1-\theta_2)}$$

Exponentiation: What if someone asked you to find $(1+j\sqrt{3})^4$? Seems like a real pain right? Not really. Rewriting in polar form, this question becomes quite manageable. Observe:

$$(1+j\sqrt{3})^4 = (2e^{j\pi/3})^4 = 2^4 e^{j4\pi/3} = 16(\cos(4\pi/3)+j\sin(4\pi/3)) = -8-8\sqrt{3}j$$

This general procedure of raising complex numbers to powers is known as DeMoivre's Theorem.

Roots: Now if you really want to impress your friends ask them, "If the square root of -1 is j, then what is the square root of j?" If that doesn't stump them, try this one: "Hey Joe, the two square roots of the number 4 are 2 and -2, right? Can you name the three cube roots of 8?" Again, the key to answering this question is rewriting the number in polar form. The other roots are evenly spaced along a circle in the complex plane. Verify for yourself that the two square roots of j are $(1+j)/\sqrt{2}$ and $-(1+j)/\sqrt{2}$.

$$\sqrt[N]{a+bj} = (re^{j\theta})^{1/N} = r^{1/N} e^{j\left(\frac{\theta}{N}+\frac{2\pi k}{N}\right)},\ k = 0, 1, \ldots, N-1$$

$$\sqrt[3]{8} = (8e^{j0})^{1/3} = 8^{1/3} e^{j\left(0+\frac{2\pi k}{3}\right)},\ k = 0, 1, 2$$

The three cube roots of 8

$-1+j\sqrt{3}$

2

$-1-j\sqrt{3}$

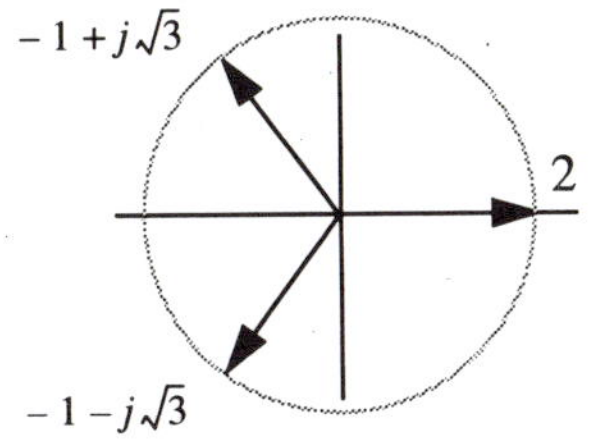

Summary

It still may not be clear to you why complex numbers are really necessary or why they have anything to do with signals and systems. The answer lies in the fact that complex exponentials are eigenfunctions of LTI systems. That probably won't make sense to you either, unless you read the rest of this book.

A.2 Basic Linear Circuit Theory

Some of the most basic linear circuit concepts are illustrated in this section. For more information, consult a introductory circuits textbook like *Engineering Circuit Analysis* by Hayt and Kemmerly (1986).

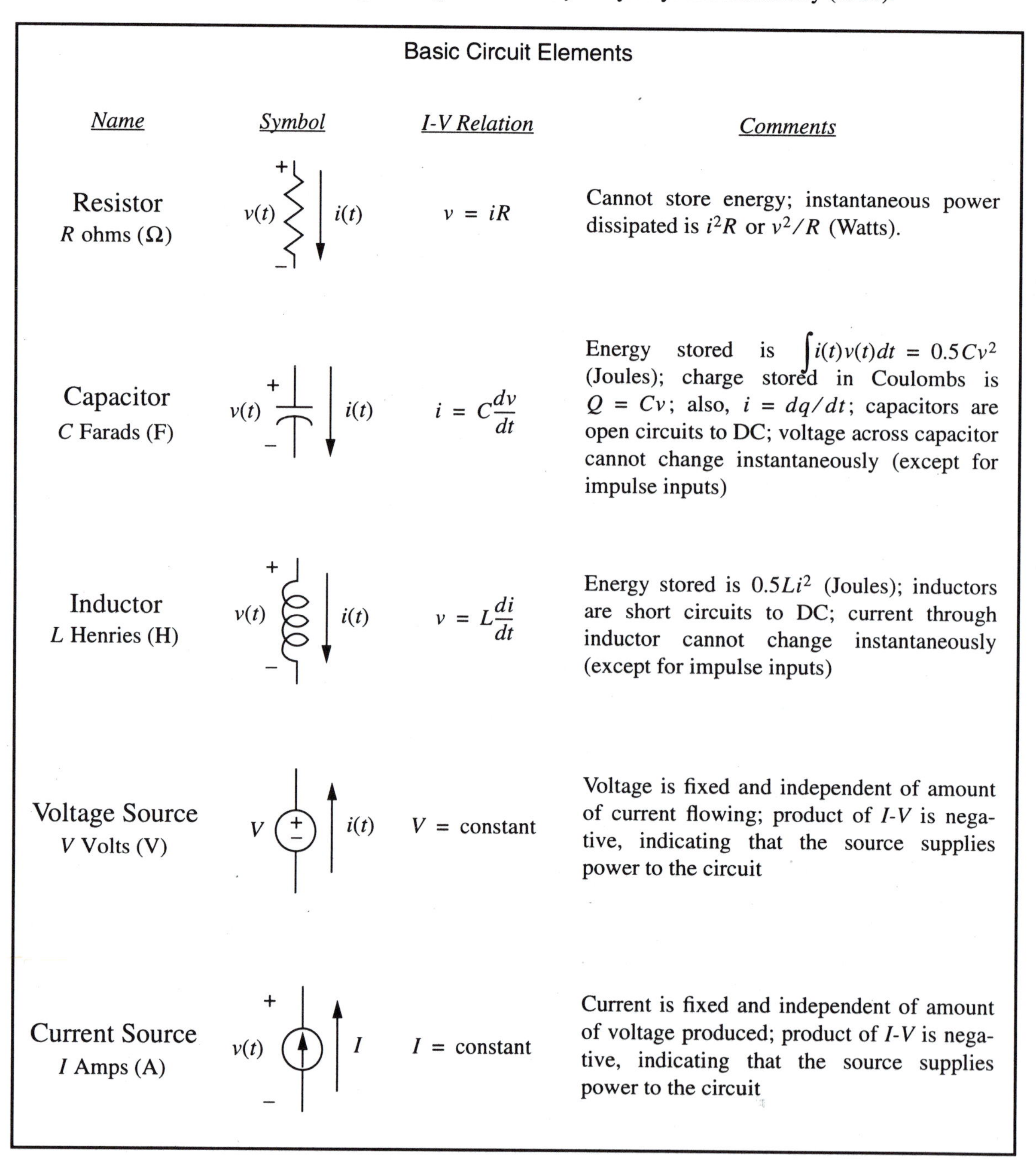

Basic Circuit Elements

Name	Symbol	I-V Relation	Comments
Resistor R ohms (Ω)	$v(t)$, $i(t)$	$v = iR$	Cannot store energy; instantaneous power dissipated is i^2R or v^2/R (Watts).
Capacitor C Farads (F)	$v(t)$, $i(t)$	$i = C\frac{dv}{dt}$	Energy stored is $\int i(t)v(t)dt = 0.5Cv^2$ (Joules); charge stored in Coulombs is $Q = Cv$; also, $i = dq/dt$; capacitors are open circuits to DC; voltage across capacitor cannot change instantaneously (except for impulse inputs)
Inductor L Henries (H)	$v(t)$, $i(t)$	$v = L\frac{di}{dt}$	Energy stored is $0.5Li^2$ (Joules); inductors are short circuits to DC; current through inductor cannot change instantaneously (except for impulse inputs)
Voltage Source V Volts (V)	V, $i(t)$	V = constant	Voltage is fixed and independent of amount of current flowing; product of I-V is negative, indicating that the source supplies power to the circuit
Current Source I Amps (A)	$v(t)$, I	I = constant	Current is fixed and independent of amount of voltage produced; product of I-V is negative, indicating that the source supplies power to the circuit

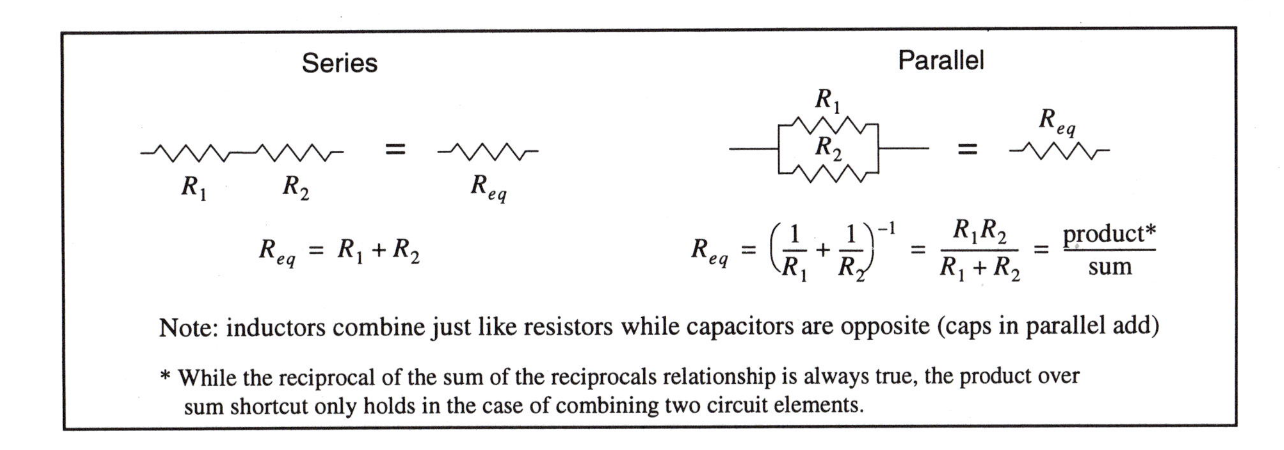

Series
Parallel
R_1
R_2
R_{eq}
$R_{eq} = R_1 + R_2$
$R_{eq} = \left(\frac{1}{R_1} + \frac{1}{R_2}\right)^{-1} = \frac{R_1 R_2}{R_1 + R_2} = \frac{\text{product*}}{\text{sum}}$
Note: inductors combine just like resistors while capacitors are opposite (caps in parallel add)
* While the reciprocal of the sum of the reciprocals relationship is always true, the product over sum shortcut only holds in the case of combining two circuit elements.

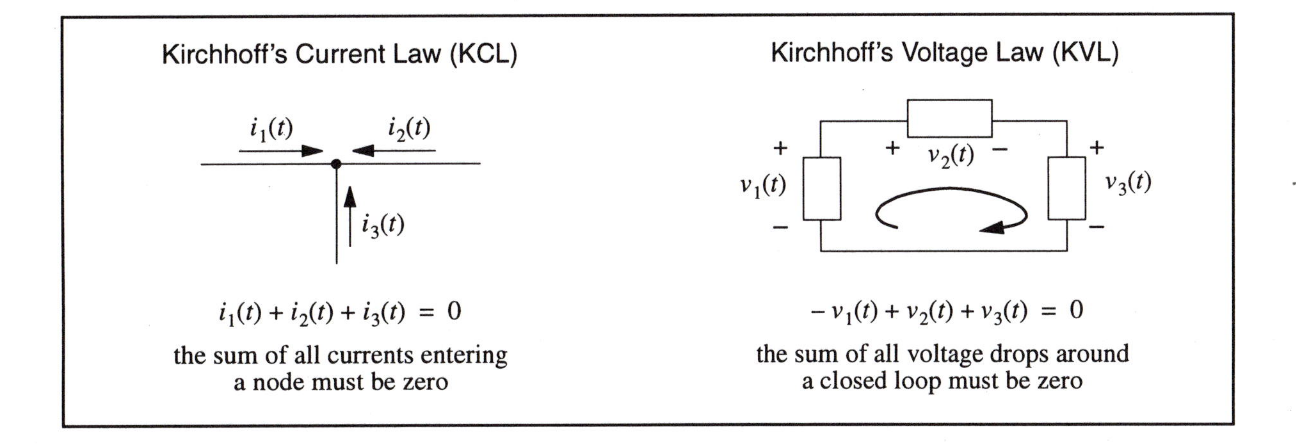

Kirchhoff's Current Law (KCL)
Kirchhoff's Voltage Law (KVL)
$i_1(t)$
$i_2(t)$
$i_3(t)$
$v_1(t)$
$v_2(t)$
$v_3(t)$
$i_1(t) + i_2(t) + i_3(t) = 0$
the sum of all currents entering a node must be zero
$-v_1(t) + v_2(t) + v_3(t) = 0$
the sum of all voltage drops around a closed loop must be zero

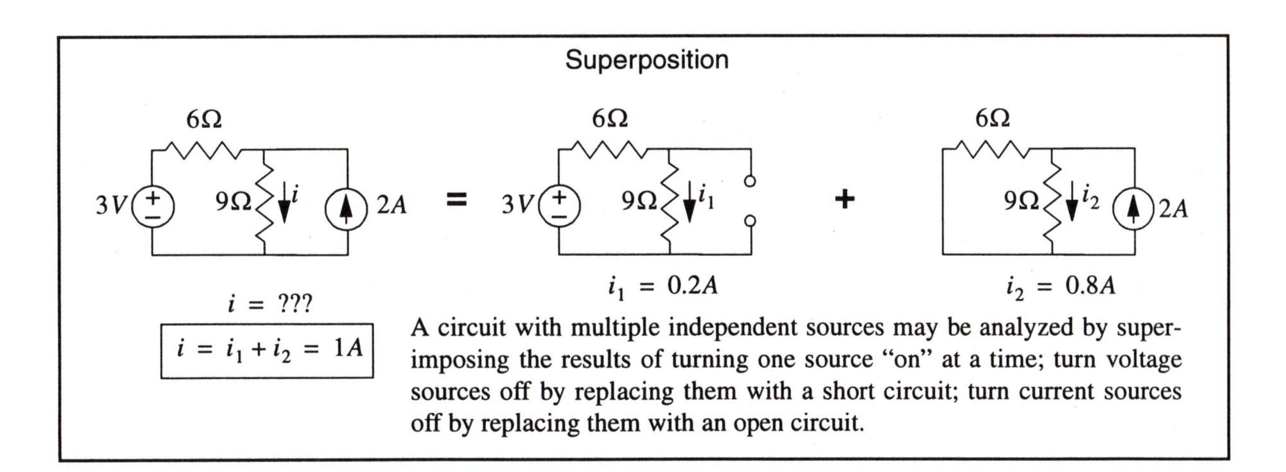

Superposition
6Ω
9Ω
3V
2A
$i = ???$
$i = i_1 + i_2 = 1A$
$i_1 = 0.2A$
$i_2 = 0.8A$
A circuit with multiple independent sources may be analyzed by superimposing the results of turning one source "on" at a time; turn voltage sources off by replacing them with a short circuit; turn current sources off by replacing them with an open circuit.

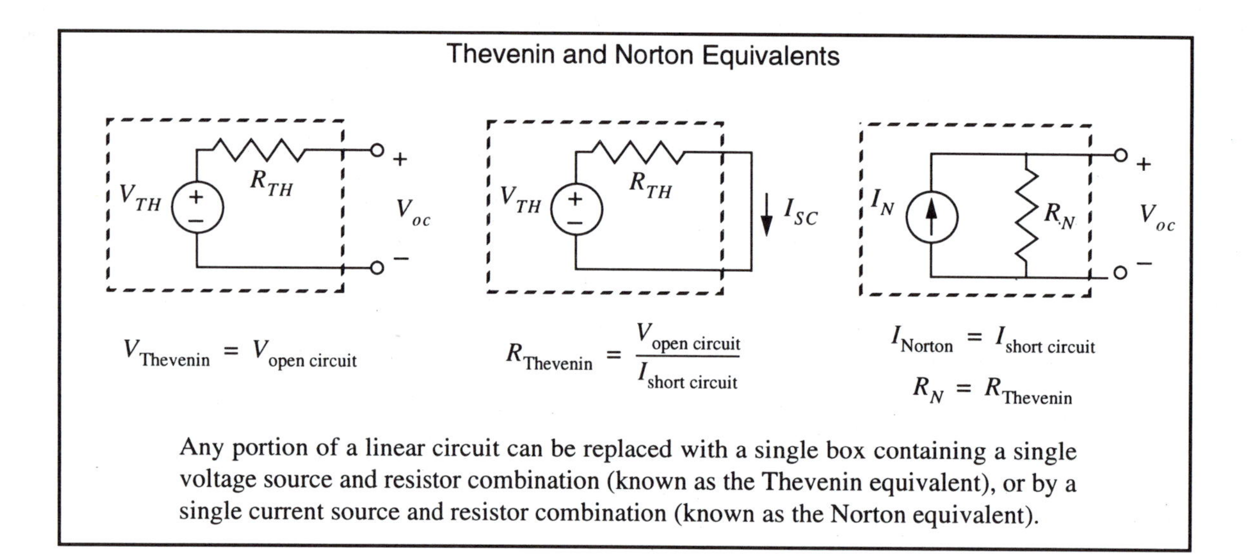
Thevenin and Norton Equivalents
V_{TH}
R_{TH}
V_{oc}
I_{SC}
I_N
R_N
$V_{\text{Thevenin}} = V_{\text{open circuit}}$
$R_{\text{Thevenin}} = \frac{V_{\text{open circuit}}}{I_{\text{short circuit}}}$
$I_{\text{Norton}} = I_{\text{short circuit}}$
$R_N = R_{\text{Thevenin}}$
Any portion of a linear circuit can be replaced with a single box containing a single voltage source and resistor combination (known as the Thevenin equivalent), or by a single current source and resistor combination (known as the Norton equivalent).

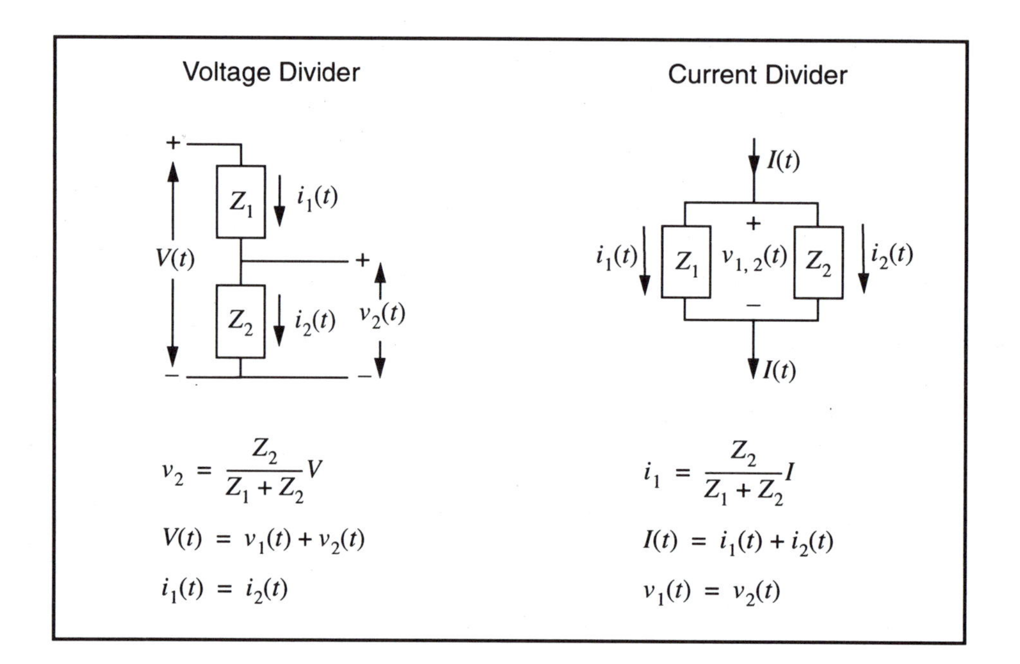
Voltage Divider
Current Divider
Z_1
Z_2
$i_1(t)$
$i_2(t)$
$V(t)$
$v_2(t)$
$I(t)$
$v_{1,2}(t)$
$v_2 = \frac{Z_2}{Z_1 + Z_2}V$
$V(t) = v_1(t) + v_2(t)$
$i_1(t) = i_2(t)$
$i_1 = \frac{Z_2}{Z_1 + Z_2}I$
$I(t) = i_1(t) + i_2(t)$
$v_1(t) = v_2(t)$

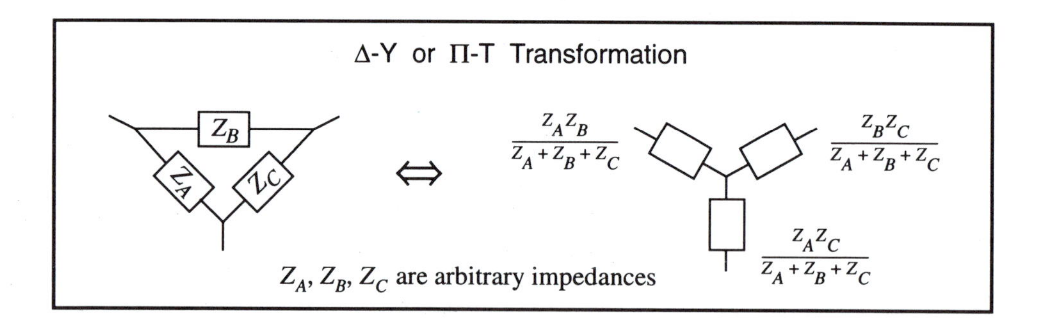
Δ-Y or Π-T Transformation
Z_B
Z_A
Z_C
$\Leftrightarrow$
$\frac{Z_A Z_B}{Z_A + Z_B + Z_C}$
$\frac{Z_B Z_C}{Z_A + Z_B + Z_C}$
$\frac{Z_A Z_C}{Z_A + Z_B + Z_C}$
Z_A, Z_B, Z_C are arbitrary impedances

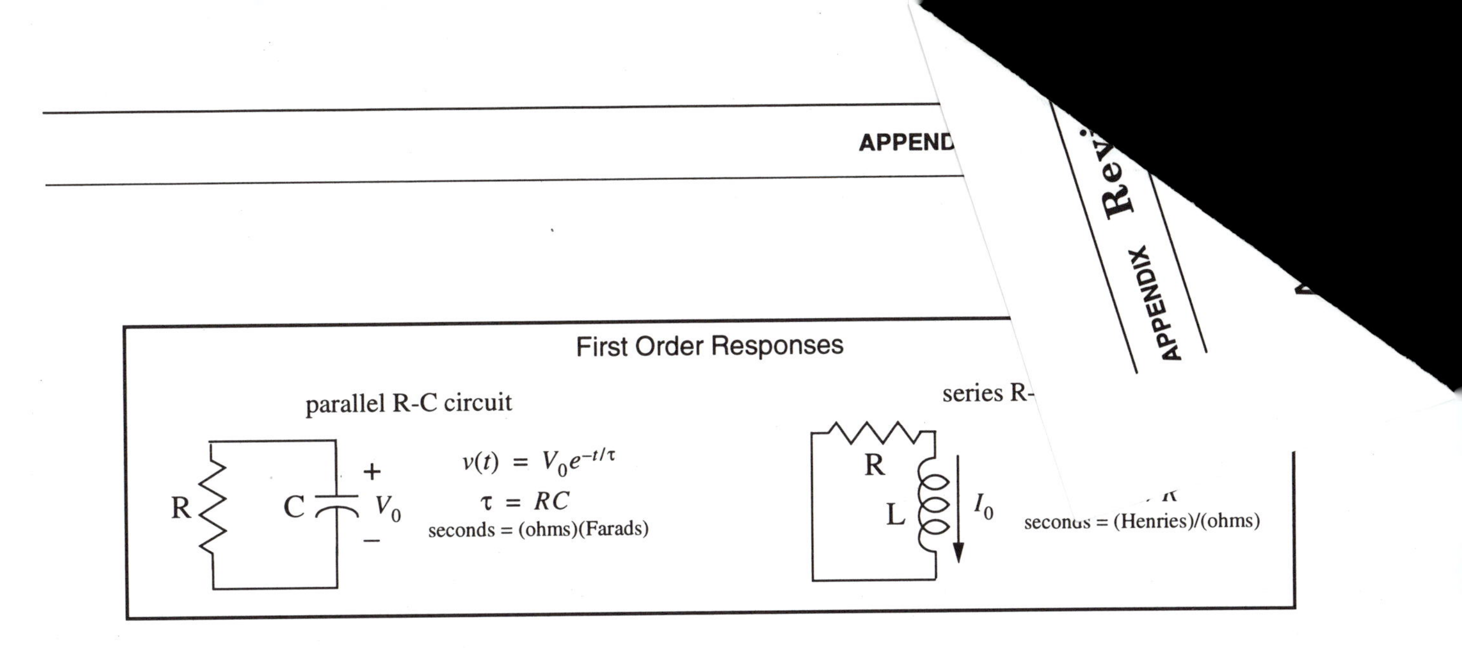

Op-Amps

An ideal operational amplifier, or op-amp for short, is nothing more than a voltage-controlled voltage source with a very high gain.

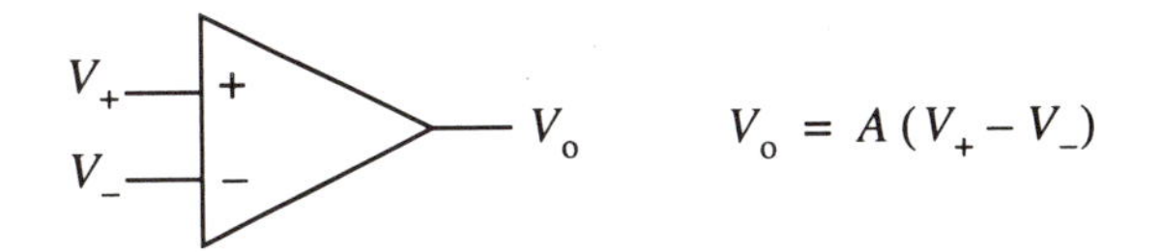

$$V_o = A(V_+ - V_-)$$

Characteristics of the ideal op-amp:

- Both the V+ and V- terminals have infinite input impedance, meaning that no current enters or leaves the op-amp there.
- The output node has zero output resistance, meaning that it behaves like an ideal voltage source supplying as much current as necessary.
- When hooked up in a feedback configuration with the output finite, the voltages at the two input terminals are identical.
- The op-amp gain A is infinitely large. Note that in real op-amps the gain A is still quite large (say around 10^6), but is never known exactly. That's why op-amps are primarily used in feedback configurations where the effects of variations in A are eliminated.

A.3 Quadratic Formula

The roots of a second order polynomial can be found by using the quadratic formula as shown below:

The two solutions of $ax^2 + bx + c = 0$ are $x = \dfrac{-b \pm \sqrt{b^2 - 4ac}}{2a}$

Sum of roots: $-(b/a)$

Product of roots: c/a

.4 Trigonometric Identities

Here are a few of the more common trigonometric identities that you may want to keep in mind. If you don't know them, at least read them over and remember that they exist so that you can look them up if necessary.

$$\sin^2\theta + \cos^2\theta = 1$$

$$\sin 2\theta = 2\sin\theta\cos\theta$$

$$\cos 2\theta = \cos^2\theta - \sin^2\theta = 2\cos^2\theta - 1 = 1 - 2\sin^2\theta$$

$$\sin\frac{\theta}{2} = \pm\sqrt{\frac{1-\cos\theta}{2}}$$

$$\cos\frac{\theta}{2} = \pm\sqrt{\frac{1+\cos\theta}{2}}$$

$$\sin(\alpha+\beta) = \sin\alpha\cos\beta + \cos\alpha\sin\beta$$

$$\sin(\alpha-\beta) = \sin\alpha\cos\beta - \cos\alpha\sin\beta$$

$$\cos(\alpha+\beta) = \cos\alpha\cos\beta - \sin\alpha\sin\beta$$

$$\cos(\alpha-\beta) = \cos\alpha\cos\beta + \sin\alpha\sin\beta$$

$$2\sin\alpha\sin\beta = \cos(\alpha-\beta) - \cos(\alpha+\beta)$$

$$2\cos\alpha\cos\beta = \cos(\alpha-\beta) + \cos(\alpha+\beta)$$

$$2\sin\alpha\cos\beta = \sin(\alpha+\beta) + \sin(\alpha-\beta)$$

$$2\cos\alpha\sin\beta = \sin(\alpha+\beta) - \sin(\alpha-\beta)$$

$$\sin\alpha + \sin\beta = 2\sin\left(\frac{\alpha+\beta}{2}\right)\cos\left(\frac{\alpha-\beta}{2}\right)$$

$$\cos\alpha + \cos\beta = 2\cos\left(\frac{\alpha+\beta}{2}\right)\cos\left(\frac{\alpha-\beta}{2}\right)$$

$$\sin\alpha - \sin\beta = 2\cos\left(\frac{\alpha+\beta}{2}\right)\sin\left(\frac{\alpha-\beta}{2}\right)$$

$$\cos\alpha - \cos\beta = -2\sin\left(\frac{\alpha+\beta}{2}\right)\sin\left(\frac{\alpha-\beta}{2}\right)$$

$$A\cos(x) + B\sin(x) = \sqrt{A^2+B^2}\cos(x - \tan^{-1}(B/A))$$

A.5 Partial Fraction Expansion

Partial fraction expansion (PFE) is the name given to the process of expanding the ratio of polynomials into the sum of ratios of smaller polynomials. The first step is to make sure that the degree of the denominator is *greater* than the degree of the numerator (not greater than or equal to). If this isn't the case, then perform long division on the offending fraction until you have something you can work with. PFE is quite useful in a variety of situations, especially when doing inverse Laplace or Z-transforms. The easiest way to describe the process is through the following examples:

Example 1: $$X(s) = \frac{2s+3}{s^2+7s+12} = \frac{2s+3}{(s+4)(s+3)} = \frac{A}{s+4} + \frac{B}{s+3} = \frac{5}{s+4} - \frac{3}{s+3}$$

Example 2: $$X(s) = \frac{2s^3}{s^2-4} = 2s + \frac{8s}{s^2-4} = 2s + \frac{A}{s-2} + \frac{B}{s+2} = 2s + \frac{4}{s-2} + \frac{4}{s+2}$$

The values for A and B in the above examples are known as the *residues* or *residuals*. How did we find these values? Well, it depends on the situation:

Linear, Non-Repeating Factors

For denominator polynomials that can be broken up into simple first-order (non-repeating) factors, solving for the residuals is actually quite easy. The general procedure is:

PFE for linear, non-repeated factors

$$X(s) = \frac{k_1}{(s-p_1)} + \frac{k_2}{(s-p_2)} + \dots$$

$$k_i = X(s)(s-p_i)\big|_{s=p_i}$$

One of the previous examples using this method:

$$X(s) = \frac{2s+3}{(s+4)(s+3)} = \frac{A}{s+4} + \frac{B}{s+3} \qquad A = \left.\frac{2s+3}{s+3}\right|_{s=-4} = 5$$

Always check your work by plugging in for A and B and recombining into one denominator!

$$B = \left.\frac{2s+3}{s+4}\right|_{s=-3} = -3$$

Linear, Repeating Factors

What about polynomials with multiple order roots? This becomes a bit more complicated.

PFE for linear repeated factors

$$X(s) = \frac{N(s)}{(s-p_1)^m G(s)} = \frac{A_0}{(s-p_1)^m} + \dots + \frac{A_k}{(s-p_1)^{m-k}} + \dots + \frac{A_{m-1}}{s-p_1} + \frac{M(s)}{G(s)}$$

$$A_k = \frac{1}{k!}\left(D_s^k\left[\frac{N(s)}{G(s)}\right]\right)\Bigg|_{s=p_1}$$

D_s^k = differentiation of the kth order with respect to s

The above formula is somewhat difficult to remember and is probably best used when programming a computer to do partial fraction expansion for you. Let's try another method for handing linear repeated factors. When finding the residual for <u>the highest power term</u> in the expansion of a repeated root, note that the above formula reduces to the standard procedure for non-repeated roots. To illustrate, let's dive into an example:

$$X(s) = \frac{1}{(s+2)^3(s+1)} = \frac{A_0}{(s+2)^3} + \frac{A_1}{(s+2)^2} + \frac{A_2}{(s+2)} + \frac{B}{s+1}$$

$$A_0 = X(s)(s+2)^3\big|_{s=-2} = -1 \qquad B = X(s)(s+1)\big|_{s=-1} = 1$$

Now, how are we going to find A_1 and A_2? One method would be: plug in for A_0 and B, recombine into one fraction; match coefficients on appropriate powers of s in the resulting numerator to the coefficients of s in the original numerator of $X(s)$; and solve the resulting system of equations for A_1 and A_2. Another method is to subtract off the highest order term from both sides, and repeat the PFE procedure until all residuals are found. The example is finished using the latter method as shown below:

subtract from both sides

$$X(s) = \frac{1}{(s+2)^3(s+1)} = \frac{-1}{(s+2)^3} + \frac{A_1}{(s+2)^2} + \frac{A_2}{(s+2)} + \frac{1}{s+1}$$

$$X_{new}(s) = \frac{1}{(s+2)^3(s+1)} - \frac{-1}{(s+2)^3}$$

$$= \frac{1}{(s+2)^3(s+1)} + \frac{s+1}{(s+2)^3(s+1)}$$

$$= \frac{1}{(s+2)^2(s+1)}$$

$$X_{new}(s) = \frac{A_1}{(s+2)^2} + \frac{A_2}{(s+2)} + \frac{1}{s+1}$$

$$A_1 = X_{new}(s)(s+2)^2\big|_{s=-2} = -1$$

repeat subtraction procedure to find A_2

Complex Roots

Sometimes factoring the denominator of $X(s)$ (with real coefficients) will lead to complex conjugate roots. In that case, the residuals are always complex conjugates of each other. Take the following example:

$$X(s) = \frac{1}{s^2+2s+2} = \frac{1}{(s+(1+j))(s+(1-j))} = \frac{A}{s+(1+j)} + \frac{A^*}{s+(1-j)}$$

$$A = X(s)(s+(1+j))\big|_{s=-1-j} = -\frac{1}{2j}$$

Nonlinear Factors

What if you don't want to factor the denominator of $X(s)$ into possibly complex roots? For second order terms, completing the square might prove useful. Otherwise, leave the higher order term as is and use a polynomial for its residual (the polynomial should be one order less than the denominator). Take the following example:

$$X(s) = \frac{25}{(s^2+2s+10)(s+5)} = \frac{As+B}{s^2+2s+10} + \frac{C}{s+5}$$

get C by standard method

$$C = X(s)(s+5)\big|_{s=-5} = 1$$

get A and B by cross multiplying and matching coefficients in numerators

$$(A+C)s^2 + (5A+B+2C)s + (5B+10C) = 25$$

$$A+C = 0 \qquad 5A+B+2C = 0 \qquad 5B+10C = 25$$

$A = -1$
$B = 3$
$C = 1$

A.6 Logarithms

The logarithm is an extremely common mathematical function that for some reason often strikes fear in the hearts of many students. Logarithms (or "logs" for short) become much easier to deal with when you remember that they are the inverse operation of exponentiation. For example, $10^3 = 1000$ and $\log_{10}1000 = 3$. Logarithms are useful for amplifying differences in very small values yet at the same time compressing differences in very large values; the difference between $\log_{10}1000$ and $\log_{10}100$ is the same as the difference between $\log_{10}0.1$ and $\log_{10}0.01$. Here is a plot illustrating the inverse relationship between exponents and logarithms:

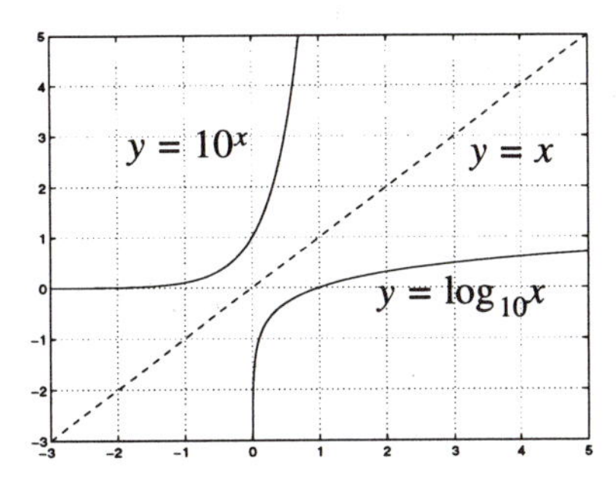

In the previous examples, the number 10 is known as the *base* of the logarithm. When the base is 10, it is often omitted, in which case the function is called the common logarithm. When the base is the transcendental number *e* (2.718281828...), the function is written as *ln* and is known as the natural logarithm. Logarithms can exist in any base, and there is a simple way of converting between them. But first, implant the following core relationship between logarithms and exponentiation firmly in your brain; use the arrows to help you remember.

How Logarithms Work

$$a^? = b \Leftrightarrow \log_a b = ?$$

Properties of Logarithms
(for any base $a>0$)

some data points	$\log_a a = 1$	$\log_a(0) = -\infty \quad (a > 1)$
	$\log_a(1) = 0$	$\log_a(\text{negative num}) = \text{undefined}$
base conversion	$\log_a b = \dfrac{\ln(b)}{\ln(a)} = \dfrac{\log(b)}{\log(a)} = \dfrac{\log_c b}{\log_c a}$	
multiplication = addition	$\log_a(xy) = \log_a x + \log_a y$	
division = subtraction	$\log_a(x/y) = \log_a x - \log_a y$	
inverse operations	$a^{\log_a x} = x$	$e^{\ln x} = x$
exponents in logs	$\log_a(b^r) = r\log_a b$	

Examples

1. Verify the following relationships:

$$\log 100 = 2 \qquad 20\log 0.001 = -60 \qquad e^{-ln2} = 0.5$$

2. Solve for x in the following equation: $2^x = 5$

$$x = \log_2 5 = \frac{\log 5}{\log 2} = \frac{0.699}{0.301} = 2.322$$

3. Evaluate $\int_0^3 2^x dx$

$$\int_0^3 2^x dx = \int_0^3 e^{ln(2^x)} dx = \int_0^3 e^{(ln2)x} dx = \frac{1}{ln2} e^{(ln2)x}\Big|_0^3 = \frac{1}{ln2} 2^x\Big|_0^3 = \frac{7}{ln2} = 10.1$$

A.7 Sequences and Series

Taylor Series

A Taylor series expansion is a polynomial of infinite length used for approximating functions around a particular operating point. The number of terms used depends on the accuracy desired in the approximation. The series expansion can be derived using the following procedure:

> If a function and all its higher order derivatives are known at point a, then the function evaluated at point x can be found by using the following formula. The closer x is to a, the fewer terms you need for accurate results.
>
> $$f(x) = \sum_{n=0}^{\infty} \frac{f^n(a)}{n!}(x-a)^n \qquad f^n(a) = n^{\text{th}} \text{ derivative of } f \text{ evaluated at } a$$

Examples

$$\frac{1}{1-x} = \sum_{n=0}^{\infty} x^n = 1 + x + x^2 + x^3 + \dots \quad \text{for } |x| < 1$$

$$\cos(x) = \sum_{n=0}^{\infty} \frac{(-1)^n x^{2n}}{(2n)!} = 1 - \frac{x^2}{2!} + \frac{x^4}{4!} - \frac{x^6}{6!} + \dots$$

$$e^x = \sum_{n=0}^{\infty} \frac{x^n}{n!} = 1 + \frac{x}{1!} + \frac{x^2}{2!} + \frac{x^3}{3!} + \dots$$

$$\sin(x) = \sum_{n=0}^{\infty} \frac{(-1)^n x^{2n+1}}{(2n+1)!} = x - \frac{x^3}{3!} + \frac{x^5}{5!} - \frac{x^7}{7!} + \dots$$

$$\ln(1+x) = \sum_{n=0}^{\infty} \frac{(-1)^n x^{n+1}}{n+1} = x - \frac{x^2}{2} + \frac{x^3}{3} - \frac{x^4}{4} + \dots \quad \text{for } |x| < 1$$

In all cases, the smaller x is, the fewer terms you need...

Other Useful Formulas

A Definition of e $$\lim_{n \to \infty}\left(1 + \frac{x}{n}\right)^n = e^x$$

Infinite Geometric Series $$a_1 \sum_{n=0}^{\infty} r^n = a_1 + a_1 r + a_1 r^2 + a_1 r^3 + \ldots = \frac{a_1}{1-r} \quad |r| < 1$$

Finite Geometric Series $$a_1 \sum_{n=0}^{N} r^n = \underbrace{a_1 + a_1 r + a_1 r^2 + \ldots + a_1 r^N}_{\text{N+1 terms}} = \frac{a_1(1-r^{N+1})}{1-r}$$

Positive Numbers $$\sum_{n=1}^{N} n = 1 + 2 + 3 + 4 + 5 + \ldots + N = \frac{N(N+1)}{2}$$

Odd Numbers $$\sum_{n=1}^{N} (2n-1) = 1 + 3 + 5 + 7 + 9 + \ldots + (2N-1) = N^2$$

Examples with Infinite Series

The rest of this section illustrates a few applications of infinite series. See how many you can solve without looking at the solutions that follow. Some of them are rather tricky!

Problems

1. Express $0.32432432\ldots$ as a fraction.

2. Evaluate $1 + 2\left(\frac{1}{2}\right) + 3\left(\frac{1}{2}\right)^2 + 4\left(\frac{1}{2}\right)^3 + \ldots$

3. Evaluate $\displaystyle\sum_{n=1}^{\infty} \frac{1}{(n+1)n}$

4. Evaluate the following repeating fraction:

$$\cfrac{2}{2 + \cfrac{2}{2 + \cfrac{2}{2 + \ldots}}} = ?$$

5. Find the equivalent resistance between A and B

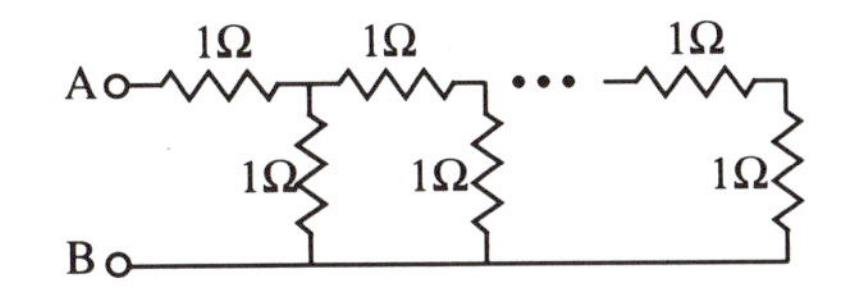

Solutions

1. $$0.\overline{324} = 324\,(10^{-3}) + 324\,(10^{-6}) + 324\,(10^{-9}) + \ldots = \frac{324\,(10^{-3})}{1-10^{-3}} = \frac{324}{999} = \boxed{\frac{12}{37}}$$

2. $$1 + \frac{1}{2} + \left(\frac{1}{2}\right)^2 + \left(\frac{1}{2}\right)^3 + \ldots = \frac{1}{1-0.5} = 2$$

$$\frac{1}{2} + \left(\frac{1}{2}\right)^2 + \left(\frac{1}{2}\right)^3 + \ldots = \frac{0.5}{1-0.5} = 1$$

$$+ \qquad \left(\frac{1}{2}\right)^2 + \left(\frac{1}{2}\right)^3 + \ldots = \frac{0.25}{1-0.5} = 0.5$$

geometric sequence

$$1 + 2\left(\frac{1}{2}\right) + 3\left(\frac{1}{2}\right)^2 + 4\left(\frac{1}{2}\right)^3 + \ldots = \frac{2}{1-0.5} = \boxed{4}$$

3. $$\sum_{n=1}^{\infty} \frac{1}{(n+1)\,n} = \sum_{n=1}^{\infty}\left(\frac{1}{n} - \frac{1}{n+1}\right) = \left(1-\frac{1}{2}\right) + \left(\frac{1}{2}-\frac{1}{3}\right) + \left(\frac{1}{3}-\frac{1}{4}\right) + \ldots + \left(\frac{1}{n}-\frac{1}{n+1}\right)$$

(cancels) (cancels)

$$= \lim_{n\to\infty}\left(1 - \frac{1}{n+1}\right) = \boxed{1}$$

4. Rewrite repeating fraction as $\dfrac{2}{2+x} = x$ $\Rightarrow x^2 + 2x - 2 = 0$

$$\Rightarrow x = -1 + \sqrt{3} = \boxed{0.732}$$

5. If x is the equivalent resistance between points A and B, then the circuit can be redrawn as:

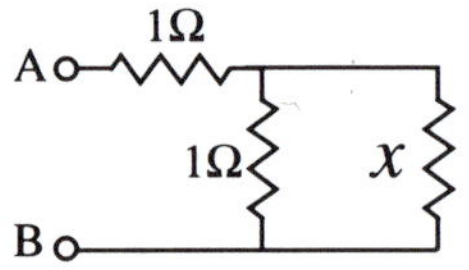

$$x = 1\Omega + 1\Omega \parallel x \qquad \Rightarrow x^2 - x - 1 = 0$$

$$\Rightarrow x = 1 + \frac{x}{x+1} \qquad \Rightarrow x = \boxed{\frac{1+\sqrt{5}}{2}}$$

(the negative solution is rejected)

A.8 Binomial Expansion

The general formula for the expansion of a binomial raised to a power is:

$$(a+b)^n = \sum_{k=0}^{n} \binom{n}{k} a^{n-k} b^k \qquad \text{where } \binom{n}{k} = \frac{n(n-1)(n-2)\dots(n-(k-1))}{k!}$$

$$= \frac{n!}{k!(n-k)!} \quad \text{if } n \text{ is a positive integer}$$

An often used form of the above formula is:

$$(1+x)^n = 1 + nx + \binom{n}{2}x^2 + \binom{n}{3}x^3 + \dots x^n$$

$$\approx 1 + nx \quad \text{if } x \text{ is small}$$

A less commonly known fact is that the power "n" does not always have to be an integer:

$$\sqrt[3]{(1+x)} = 1 + (1/3)x + \frac{(1/3)(-2/3)}{2!}x^2 + \frac{(1/3)(-2/3)(-5/3)}{3!}x^3 + \dots$$

A.9 Linear Algebra

Linear algebra facilitates the analysis of large linear time-invariant systems and becomes practically mandatory when dealing with multiple-input, multiple-output systems. Matrices provide a useful, compact representation for the state equations of a system. Once a system is in matrix or "state-space" form, there are a wide variety of system properties that can be derived directly from the state transition matrix. Another advantage is that once in matrix form, systems can be easily analyzed using a computer software package such as MATLAB.

Since this book is primarily concerned with single-input, single-output systems, a thorough review of linear algebra will be left for another book. In the meantime, it is suggested that the reader refer to a linear algebra textbook such as *Introduction to Linear Algebra* by Strang (1993) if necessary.

Bibliography

Gene F. Franklin, J.D. Powell, and A. Emami-Naeini. *Feedback Control of Dynamic Systems*. 3rd ed. Reading, MA: Addison-Wesley, 1994.

William H. Hayt and J.E. Kemmerly. *Engineering Circuit Analysis*. 4th ed. New York: McGraw-Hill, 1986.

Paul Horowitz and W. Hill. *The Art of Electronics*. 2nd ed. New York: Cambridge University Press, 1989.

E.J. Kennedy. *Operational Amplifier Circuits: Theory and Applications*. New York: Holt, Rinehart and Winston, 1988.

Huibert Kwakernaak and R. Sivan. *Modern Signals and Systems*. Englewood Cliffs, NJ: Prentice-Hall, 1991.

Jae S. Lim. *Two-Dimensional Signal and Image Processing*. Englewood Cliffs, NJ: Prentice-Hall, 1990.

Alan V. Oppenheim, A.S. Willsky, and I. Young. *Signals and Systems*. Englewood Cliffs, NJ: Prentice-Hall, 1983.

Alan V. Oppenheim and R.W. Schafer. *Discrete-Time Signal Processing*. Englewood Cliffs, NJ: Prentice-Hall, 1989.

William McC. Siebert. *Circuits, Signals, and Systems*. New York: McGraw-Hill, 1986.

Gilbert Strang. *Introduction to Linear Algebra*. Wellesley, MA: Wellesley-Cambridge Press, 1993.

More Comments on Book
and
Publisher Contact/Ordering Information
on Back Page

More comments on Signals and Systems Made Ridiculously Simple!

ISBN 0-9643752-1-4

"In an age when many technical authors feel they must inundate their readers with thousand page textbooks, it was a pleasure to read your beautiful book. It succinctly gets to the important topics and helps the reader separate the 'wheat from the chaff'. The illustrations and typographical layout are great!"

- Senior Principal Engineer, The Boeing Company

"Finally I have a book on Signals and Systems that is written in plain English!"

- Student, University of Hawaii

"Really enjoyed your book. It helped me tie the concepts together after a long period of using some and letting the others slip by the wayside."

- Researcher, Scripps Institution of Oceanography

"I can say this is the best book I have seen which summarizes a lot of the fundamentals you really should know in a nice, clearly formatted, condensed form."

- Graduate Student, Tufts University

"A worthwhile companion to traditional texts in the field."

- Instructor, Harvard Medical School

"I was very impressed with the quality and scope of your book; it is one of the best engineering textbooks I have purchased."

- Professional Engineer, American Water Works

"This book does a really nice job explaining the difficult concepts of an introductory signals/systems course. Clearly written, with excellent diagrams, this book is a must for anyone studying signals at an undergraduate level."

- Student, Buffalo, NY

Complete ordering information is on our web site!
http://www.zizipress.com

For further information, please write, fax, e-mail, or call!

ZiZi Press
1404 Old Carriage Lane
Huntsville, AL 35802
USA

Tel: (256) 520-5249
FAX: (256) 883-3124
E-mail: books@zizipress.com
Author: zzkaru@alum.mit.edu